Study of Nuclear and Alternative Energy Systems

SUPPORTING PAPER 2

ENERGY MODELING FOR AN UNCERTAIN FUTURE

The Report of the
Modeling Resource Group
Synthesis Panel
of the
Committee on Nuclear and Alternative Energy Systems
National Research Council

NATIONAL ACADEMY OF SCIENCES
WASHINGTON, D.C. 1978

International Standard Book Number 0-309-02781-0

Library of Congress Catalog Card Number 78-60404

Available from:

Printing and Publishing Office
National Academy of Sciences
2101 Constitution Avenue, N.W.
Washington, D.C. 20418

Printed in the United States of America

A GUIDE TO THE READER

The following key figures and tables and the text of the sections in which they appear may give a quick indication of the methods and principal findings of this report.

Figure II.1, Section II.2, page 9

Table III.1, Section III.1, page 37

Other key sections and tables in the report are:

Section III.2

Table III.20, Section III.8, page 100

Table III.23, Section III.9, page 110

Finally, Section IV.2 outlines the methodology of the MRG analysis of research, development, and demonstration programs under uncertainty.

PERSONS CONNECTED WITH THE MODELING RESOURCE GROUP

MEMBERS OF THE MODELING RESOURCE GROUP

Tjalling C. Koopmans *(Chairman)*, Cowles Foundation, Yale University
Edward G. Cazalet, Private Consultant
Kenneth C. Hoffman, Brookhaven National Laboratory
William W. Hogan, Stanford University
Michael Kennedy, University of Texas at Austin
Lester B. Lave, Carnegie-Mellon University
Alan S. Manne, Stanford University
William D. Nordhaus, Council of Economic Advisors
Philip K. Verleger, Department of the Treasury
David O. Wood, Massachusetts Institute of Technology

MEMBERS OF THE RESEARCH AND DEVELOPMENT SUBGROUP

Kenneth C. Hoffman *(Chairman)*, Brookhaven National Laboratory
Edward G. Cazalet, Private Consultant
Harry Davitian, Brookhaven National Laboratory
Alan S. Manne, Stanford University
William D. Nordhaus, Council of Economic Advisors
Richard Richels, Electric Power Research Institute

ASSOCIATE MEMBERS OF MRG

William Marcuse, Brookhaven National Laboratory
David Nissen, Federal Energy Administration

CONSULTANTS TO THE MODELING RESOURCE GROUP

DeVerle P. Harris, Arizona University
Robert E. Litan, Council of Economic Advisors
James L. Sweeney, Stanford University
Lester D. Taylor, Arizona University
John P. Weyant, Stanford University

PREFACE

In June 1975, the National Research Council (NRC) undertook a comprehensive study of the nation's prospective energy economy during the period 1985-2010, with special attention to the role of nuclear power among the alternative energy systems. The goal of the study is to assist the American people and government in formulating energy policy.

The Governing Board of the National Research Council appointed an NRC-wide Committee on Nuclear and Alternative Energy Systems (CONAES) to conduct the study. CONAES consists of 15 members drawn from diverse disciplines and backgrounds. The committee developed a three-tiered functional structure for the study. The first tier is CONAES itself. The ultimate findings, judgments, and conclusions of the study will be embodied in its final report.

To provide scientific and engineering data and analyses, a second tier of four panels was formed to examine (1) energy demand and conservation, (2) energy supply and delivery systems, (3) risks and impacts of energy supply and use, and (4) syntheses of diverse models of future energy economies, respectively. Each panel, in turn, established a number of resource groups--22 in all--as the third tier, to address in detail an array of more particular matters, such as buildings and transportation systems, solar energy, breeder reactors, coal technologies, health and environmental implications, and alternative consumption patterns and economic models. In all, more than 200 informed individuals served on or contributed to the work of the panels and resource groups.

The National Research Council customarily publishes only the final reports of its committees--and then only after the report has been reviewed by a group other than its authors according to procedures approved by a Report Review Committee consisting of members of the National Academy of Sciences, the National Academy of Engineering, and the Institute of Medicine. However, because such a large volume of information and analyses was assembled for consideration by the committee, and because of the diversity and scope of that information and the accompanying judgments, the panel reports and approximately 10 reports by the resource groups are being published as supporting papers.

Each of these has been considered and used by CONAES but has not undergone the critical review procedure normal to the NRC.

It must be recognized that some conclusions of the panel and resource group reports may be at variance with the conclusions of the CONAES report. The findings reported in these documents are those of their authors and are not necessarily endorsed by CONAES or the National Research Council.

This report covers the work of the Modeling Resource Group of the Synthesis Panel. It was intended to inform the panel and CONAES in their deliberations. It is published, among the other supporting papers, to chart a selection of alternative energy futures in the form of "if . . . , then . . . " statements, and to demonstrate the variety of techniques and fields of knowledge on which CONAES has had to draw.

ACKNOWLEDGMENTS

Successive drafts of Sections I, II, and III and Appendixes A through E were prepared by Tjalling Koopmans, Robert Litan, and John Weyant. Successive drafts of Section IV and Appendix F were prepared by Harry Davitian, Kenneth E. Hoffman, Alan S. Manne, William D. Nordhaus, and Richard Richels. Valuable comments and written contributions were received from Lester Lave, from members of the MRG, from other participants on the CONAES study outside the MRG, and from Jim Sweeney as a referee. Editing and proofreading were done by Daniel Preston, typing by Glena Ames and Lydia Zimmerman, and figures drawn by Glena Ames. Our thanks to all.

CONTENTS

TABLES

FIGURES

I OPTIONS, CHOICES, AND THEIR PROBABLE OUTCOMES

I.1 OPTIONS

The United States is faced with the necessity of choosing what combination of energy sources and technologies can best satisfy its energy needs over an extended future. The following list contains a number of important options:

conservation measures that may be mandated, urged, made possible by technical progress, or stimulated by increasing energy prices;
imports of oil and gas at world market prices;
coal, for process heat, electricity generation, liquefaction, or gasification;
nuclear fission, using reactors currently in operation or more advanced reactors (converters and/or breeders);
(direct) solar energy, decentralized, or as central-station electric energy;
nuclear fusion, if and when feasible;
shale oil;
other sources (biomass, geothermal, etc.).

These options can often supplement, and in some uses substitute for, each other. The choice is therefore not one between two or more mutually exclusive alternatives. Rather, the choice is that of the best mix of several options, at any time, and as it changes over time.

Taking the long view, the real cost of production of domestic oil and gas is expected to increase through depletion--relative to alternative sources of energy. We must therefore expect a period of transition in which other sources gradually take over from oil and gas, even if total U.S. energy use should remain at the present level. If historical trends of population and income per capita continue, however, there is likely to be a further increase in total energy use, which is limited by the increase in energy cost and prices it calls forth, by measures taken to protect the environment, and by efforts to limit dependence on foreign sources.

I.2 POLICY CHOICES

It helps to visualize how one could choose a good (changing) mix of these options *if* three conditions were satisfied that do not in fact prevail:

1. the success and the economic costs of development and application of these technologies at various scales could be accurately foreseen;
2. the future course of population, of per capita income, of labor force participation, of employment and productivity, and of other behavioral driving variables could be foreseen; and
3. the environmental impacts and other risks arising from given levels of use of each of these technologies could be accurately assessed in real dollar terms.

Then, for a market economy, one would ideally project paths for the use of each technology over time such that industry, transport, and final consumers would want to buy the projected quantities at prices that reflect minimum economic and environmental cost. Such a calculation--an application of "deterministic modeling"--would be mathematically complex but conceptually simple. It is one example of a *scenario,* that is, a detailed *projection* that traces the implications of explicitly stated assumptions, for the future course of production, imports, conversion, possibly transportation, and consumption of the various forms of energy.

The procedures followed by the Modeling Resource Group (MRG), described in Chapters II and III, are designed to cope with the fact that none of the three conditions listed above is satisfied. To face the fact that neither the *realizations* of (a) future technology nor those of (b) future human behavior are fully foreseeable, the MRG adopted "central" or "Base-Case" values for each uncertain technological or behavioral variable and one or two "high" and/or "low" alternative values for those variables that stand out by being both uncertain and important to the answers found. These estimates are determined on a judgmental basis and are described in Section II.2 and Appendix B. These estimates serve as explicitly stated assumptions on which the corresponding projections (or scenarios) rest. Pairwise comparison of scenarios that differ only in the value given to just one realization variable will give an estimate of the effect of the uncertainty attached to that variable.

Even so, the list of realization variables does not cover all important sources of uncertainty. In particular, we regard the fact that we do not have good measures of (c) environmental impacts as the most serious limitation of the work described in this report. What we have been able to do is to trace the economic effects of policies that limit or proscribe the use of one or more technologies with environmental or other effects or risks known to be adverse--such as coal, shale oil, and, to varying degrees, the processes of nuclear fission. This procedure makes it possible to estimate the economic costs of these policies. These costs will then have to be compared with, and balanced against, the value of the protection the policies provide, on a judgmental basis.

The MRG has not been in a position to take that last and most important step.

Mathematically and computationally, scenario comparisons that trace the effects of changes in a policy variable are not different from those that trace the effects of changes in a realization variable. Their place in the analysis is different, however, and we have kept the two sets of variables as distinct as we could in reporting on our work. We therefore use the term *policy impact analysis* when the variable that is tentatively changed is a policy variable, and the term *sensitivity analysis* when the variable that is changed is a realization variable. Important examples of policy impact analysis are the estimations of the effects of a moratorium on nuclear construction starts (nuclear "moratorium") or of limits on the use of coal, reported on in Chapter III (see Tables III.6, III.7, and III.8).

Finally, we may study how the result of a policy impact analysis itself depends on the value attributed to a realization variable. This is called *sensitivity analysis*. An example is the sensitivity of the estimated effects of a "nuclear moratorium" and/or of the placing of limits on the use of coal and shale *to* the assumed growth rate of the Gross National Product (GNP)(see Chapter III, Tables III.4 and III.8).

I.3 ENERGY MODELS

The principal tool for the estimation of internally consistent and mutually comparable scenarios, of policy impacts, and of their sensitivity to the values of realization variables is the construction of models of the energy sector of the economy. An *energy model* connects a number of energy variables with various driving variables. The simplest model would describe how total energy use and the energy price can arise from an equilibrium between a force on the demand side (say, consumers' purchasing power) and a constraint on the supply side (say, the cost of production of energy). More detailed models would distinguish several end uses of energy (heating, cooling, transportation) and several fuels, each with their own production constraints. For a determinate model, the number of relations and of binding constraints must at any time equal the number of energy variables to be projected. The relations must be estimated on the basis of empirical information--a procedure that ultimately rests also on more general assumptions about production possibilities and economic motivation and behavior. Principal empirical sources for these estimates are technological information relating to past or future production and utilization possibilities, and econometric estimation of past demand, supply, or other economic behavior relations. Much of the technological information consists of physical quantities of inputs and outputs to the various processes, or ratios of such inputs to outputs, with the input quantities sometimes aggregated into cost estimates. Most of the econometric information consists of consumers' and producers' behavior relations that connect quantities with prices. For all types of information, both the role of expert judgment in arriving at the quantitative relations and the uncertainties associated with these relations increase as one looks further into the future.

In the economy, past, present, and future are connected by changes in productive capacities through capital construction and retirement, in resource availability by gradual depletion of the more accessible deposits, and in economic behavior by gradual changes in tastes and life-styles, in size and composition of the population, and in changing regulatory and taxation policies. Mathematically speaking, the model is represented by a large system of equations or inequalities, many linear and some nonlinear. Some of the equations connect variables referring to different points in time. However, there is considerable similarity between subsystems of the equations relating to successive periods.

Appendix A describes in detail the six models that have been used in the CONAES study. Two of these, DESOM (developed at Brookhaven Laboratory by William Marcuse, extending an earlier model, BESOM, developed by Kenneth Hoffman) and SRI (Stanford Research Institute, a model developed by Edward Cazalet) describe the delivery of energy services to society with primary emphasis on the supply side of the energy sector. One, the Energy Model of Data Resources, Inc. (DRI), developed by Philip Verleger, concentrates on the demand side. The other three represent the demand and the supply sides of the energy sector in comparable detail, in a contemporal and intertemporal equilibrium of prices and quantities such that, but for stated exceptions, demand at those prices equals supply at those prices in all regions and periods.

There are two main methods in use for computing the equilibrating quantities and prices. A version of the first method is used in the Project Independence Evaluation System (PIES) model developed at the Federal Energy Administration (FEA) by William Hogan and collaborators.* This method solves the equation system by an iterative procedure of successive approximations going back and forth between quantities and prices--filling in also the quantities directly related only to other quantities--until the procedure stabilizes at a solution.

The other method is used by the two remaining, and more aggregative, equilibrium models: Energy Technology Assessment (ETA) by Alan Manne and a simulation model by William Nordhaus. This method, following a long tradition in economic thought, which in turn has precedence in classical mechanics, simulates a moving price-quantity equilibrium over time by the application of a suitable maximization or minimization principle. The method leans on an assumption of competitive markets, as a first approximation, in extraction, transportation, conversion, and use of energy in its various forms. The quantity maximized can be interpreted as the sum of discounted future utilities (benefits, satisfactions) experienced by the ultimate consumers of end uses of energy and of the energy component in the value of other consumption goods produced with energy inputs. The utility function imputed to consumers is derived from econometric estimates of the response of energy demand to energy prices.

Either of the two methods can also be applied to models that concentrate on one or the other side of the market. For instance, DESOM uses minimization of the sum of discounted costs to be incurred to meet

*David Nissen, James Sweeney, and others.

a given, postulated path of energy end uses over the entire period to 2010. On the other hand, SRI determines the allocation of primary energy from extraction through transportation, conversion, and assignment to given end uses by iterative approximation of an equilibrium on the supply side. Verleger similarly calculates an equilibrium on the demand side where the supply side is largely represented by given prices rather than quantities.

I.4 OPTIMIZATION AS A MEANS OF SIMULATION

As mentioned above, three of the models (DESOM, ETA, and Nordhaus) use the mathematical techniques of *optimization*--a term that covers both maximization of benefits, or minimization of costs, or maximization of the excess of benefits over costs, as well as the maximization of any other *objective function* that is adopted. The use of optimization in these models should be seen as a means of simulating, as a first approximation, the behavior of a system of interacting competitive markets.

The term *objective function,* too firmly established in the jargon to avoid it here, may give rise to misunderstanding. The word *objective* here is a noun used as an adjective. It refers to the goal, or "objective," pursued by the consumer. The benefits term in a goal function, or objective function, estimates the value attached by the consumer to the amount he consumes. Like many values, it is "subjective" rather than "objective" (the adjective is meant here). But it is revealed by the consumer's response to price--a circumstance that permits its estimation.*

We emphasize two features and one caveat with respect to the use made of the method of optimization in this report.

First, for suitably chosen objective functions that are expressed in constant dollar equivalents, the maximum attained value of the objective function measures the contribution to aggregate economic welfare made by the energy technologies included in the model. The economic impact of the addition or removal of any particular technology is then measured by the difference in the attainable maximum with and without that technology. This procedure will be used in Section III.8, Table III.20.

The second feature extends the first for models that take the form of *linear programming,* or some of its generalizations in *mathematical programming*. In these models the objective function and the several constraints may be linear, or suitably nonlinear.

The feature referred to consists of the automatic computation, along with the "optimal" quantity path for the energy sector, of *shadow prices*, one associated with each constraint of the model.† The two paths, one

*For further explanation, see Section III.8.

†Technically the shadow price for each resource, capital good, intermediate flow or end use of energy is, barring mathematical accidents, the derivative of the attained maximum with respect to the availability of that good from outside the system--as expressed by the constrains in question.

of "optimal" quantities and one of shadow prices are the counterpart, for models using optimization, of the quantity-price projections of those equilibrium models that do not use optimization techniques. The shadow price concept, however, also applies to situations where market prices are hard to visualize. In these cases, the shadow price concept can be of help in facing the fact that the objective functions described above usually take into account only the benefits from goods and services bought by the consumer and neglect the bads from environmental damage for which the individuals affected cannot exact compensation from the polluter. The shadow price associated with a constraint that curtails or eliminates the use of a technology measures the economic cost of that constraint. For instance, a shadow price associated with an environmental standard for maximum allowable concentration of CO_2 in the atmosphere can be used to cost out the market-economic effects of such a standard.*

A caveat is needed in situations where the benefit comparison between two future technologies, or between the presence and absence of one technology, comes out as a "close call" for one alternative against the other. In such a situation, models with linear constraints may flip-flop, in the sense that a small change in an already somewhat uncertain cost estimate may drastically change the recommendation. In such cases one must be careful in the interpretation of one's findings, especially when important environmental effects or other externalities are not recognized in the model. Signals that will alert the model user to close-call situations include the impact analysis of the arbitrary exclusion of one or the other technology. Shadow prices can also be employed as such signals.

I.5 UNCERTAINTY AND DECISION-TREE ANALYSIS

The manner in which sensitivity analysis seeks to cope with uncertainty about future realizations of technology or behavior is to project future paths of the energy sector for two or more alternative values of some or each of these driving variables. Such calculations are reported in Chapter III. Inferences from these calculations can be strengthened if it is possible, on a judgmental basis, to assign a probability distribution to the realizations for each such variable. Instances of this type of reasoning occur in Chapter IV.

However, uncertainty about future realizations, and for that matter about the acceptability of policies, diminishes as time goes on. Therefore, instead of comparing alternative future *paths* of the energy economy, defined immutably from the present forward, one might want to compare *strategies*. That is, one might want to chart alternative courses of decision making where each potential new piece of technological or behavioral information that comes along in the future leads in a preselected

*See W. Nordhaus, "Strategies for the Control of Carbon Dioxide," Cowles Foundation Discussion Paper No. 443, January 6, 1977.

way to a new course of action--which may be either a revision or a continuation of the course followed up to that point in time. Modeling according to this concept is called *decision-tree analysis*. It is particularly suited for the modeling of research, development, and demonstration (RD&D) decisions and their revision in the light of experience.

In Chapter IV we report on work done along these lines by a subcommittee of the MRG called the R&D Programming Subgroup (R&DPSG), chaired by Kenneth Hoffman. This group has sorted realization variables according to the future year (1986, 1996, . . .) by which their values are assumed to become known. Prior to its resolution date, the uncertainty about each such variable is represented by an appropriate judgmental probability distribution. As demonstration of the use of this technique, Chapter IV presents an application to the RD&D decisions with regard to advanced nuclear fission technologies.

There is no implication in this choice of application that other R&D objectives, such as solar energy, synthetic fuels, or *in situ* fuel extraction are less important than the nuclear alternatives investigated. Under constraints of time and computation expense, the R&DPSG chose to concentrate on a manageable set of alternatives that are important in current energy policy discussion.

II THE WORK OF THE MODELING RESOURCE GROUP

II.1 AIMS AND STEPS

There are two principal aims of the MRG studies. One is the estimation of impacts of changes in certain policy and realization variables on a number of important outcome variables. The latter include total production and consumption of primary energy equivalents, future mixes of supply and demand, and energy prices. The estimation of such variables also allowed the MRG to provide some answers to the important question of to what extent the GNP, or along with GNP some better measure of national welfare, is dependent on continued high levels of energy consumption. In short, the first task the MRG set for itself was to provide plausible and interesting projections of our broad energy future under the different realizations and policies to be outlined shortly. This includes the estimation of the economic value of certain energy technologies and combinations thereof, reported in Chapter III.

The second aim, pursued in Chapter IV, is a natural extension of the effort to estimate the economic value of certain energy technologies. Specifically it is to examine in a systematic way the economic desirability of government-funded research and development of certain energy technologies, in the presence of uncertainty, incorporating responses to new information about the realizations of uncertain variables and estimates of the best target dates for completion of the research and development projects. The procedure for that work differed in some ways from the method employed to accomplish the first aim, and those differences, where significant, are described in Chapter IV.

Because of the novelty of the task of comparing projections based on six energy models for time periods up to 2010, there was considerable experimentation throughout the first half of the study. As a result, a number of simulations were made on sets of assumptions that differ somewhat from the final set of assumptions defined here in Chapter II. Considerable insight was gained from those earlier simulations and, where relevant, qualitative results from such runs will be reported. These will be accompanied by a discussion of possibly important differences in underlying assumptions between these and the final set of simulations.

The steps outlined below pertain to the first task of the MRG, namely, the construction of different energy consumption and supply mix scenarios, and the economic evaluation of alternative technologies. The results are reported in Chapter III.

II.2 STEP 1: LIST OF DRIVING VARIABLES WITH BASE-CASE AND ALTERNATIVE VALUES

For conceptual clarity, the types of driving variables considered by the MRG were placed in three groups: (a) "policy" variables, (b) "realization" variables (values of which will be revealed in stages in the future), and (c) blends of (a) and (b). A given set of values for the driving variables (arising from alternative policy choices, realization outcomes, and blends of the two), then, constitutes a scenario that can serve as the input to any of the six energy models used by the MRG. Whereas variation of policy variables allows estimates to be made of policy impacts, additional variation of the values assigned to realization variables permits sensitivity analyses that trace the differences in policy impacts when alternative facts of nature or behavior pertain.

Figure II.1 provides a compressed view of how the various driving variables serve as inputs into the energy models to yield the simulation outputs. The remainder of this section lists and describes in greater detail each of the more important "driving variables" identified by the MRG in each of the three categories. In some cases, it was impossible

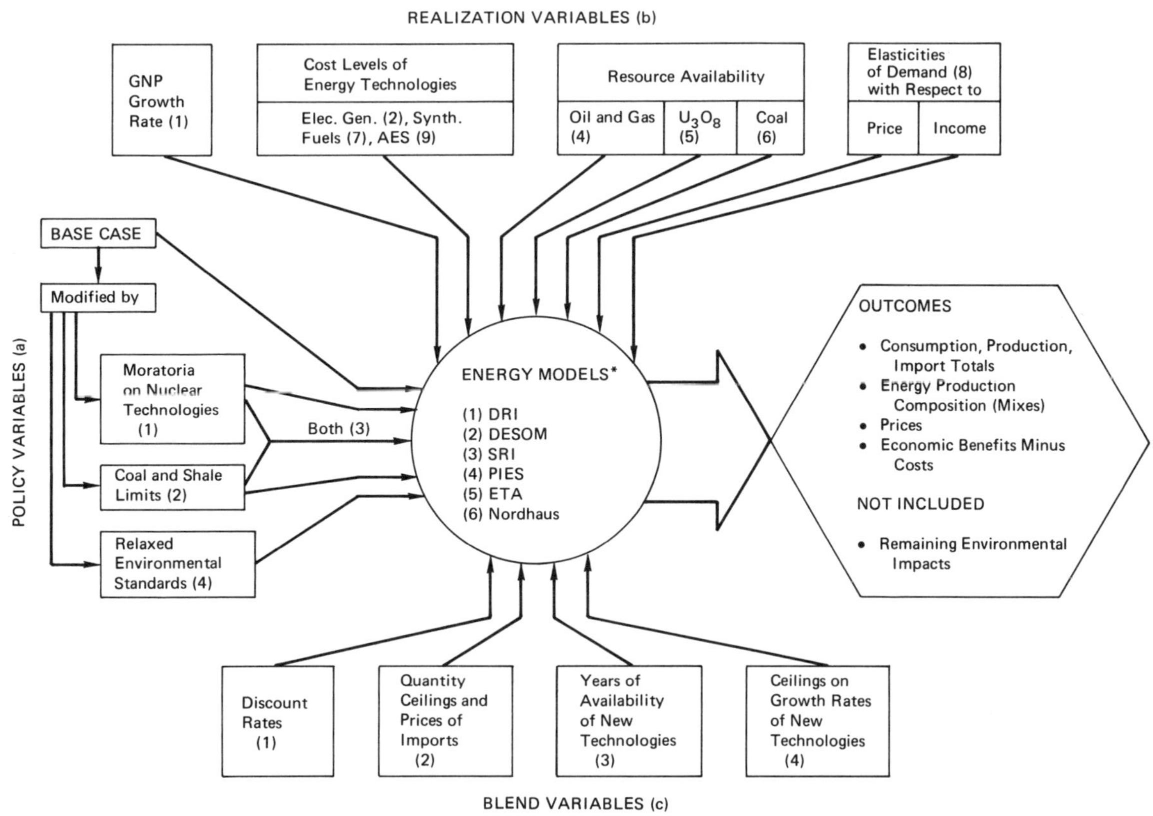

FIGURE II.1 Compressed view of driving variables, energy models, and their outcomes. *The models vary in the detail and composition of the information inputs they require and the information outputs they produce.

neatly to classify a variable, and therefore it was placed in the category felt to be, on balance, the most appropriate. Sections II.3 through II.5 describe some of the major alternative combinations of driving variables utilized in order to conduct policy impact and sensitivity analyses within and across the models. Finally, the description and analysis of the actual model simulations are given in Chapters III and IV.

a. Policy Variables

We now list and explain the "pure" policy variables that the MRG used to define alternative scenarios. We do not attempt to assign central, high, and low values to the policy variables. It would be prejudging the issue to use the probability of adoption of a certain policy as an input to a study that is intended to help in evaluating policy alternatives and choosing from among them.

1. Moratoria on development and new construction of:
   a. Fast Breeder Reactor (FBR)
   b. Advanced Converter Reactors (ACR's)
   c. All nuclear reactors
2. Limits on the annual production of coal and shale oil:
   a. "Medium" production limit--40 Quads* of coal production in 2000, ultimately approaching 50 Quads, and 2 Quads of shale oil in 2000, ultimately approaching 12 Quads.
   b. "Low" production limit--Upper limit of 25 Quads of coal and shale oil production combined.
3. The combination of nos. 1 and 2.
4. Relaxation of present environmental standards applicable to coal and nuclear technologies.
5. Non-price-induced conservation measures (10 percent non-price-induced reduction in aggregate energy consumption by 2010).

The first three policies represent alternative responses society may have to a wide set of "policy concerns," including but not limited to the environmental damages and risks associated with coal, shale oil, and nuclear technologies, and the possible risks of proliferation or diversion of weapons-grade nuclear materials. Thus, the use of nuclear technologies may be dealt with, as in policy no. 1, by imposing moratoria of varying degrees of restrictiveness, ranging from moratoria on the development of only the FBR to moratoria on the further development and construction of all ACR's,† or indeed to moratoria on the new

*One Quad corresponds to a quadrillion (10^{15}) BTU of energy content.

†The MRG defined in its Base Case a "representative ACR" to be one with a conversion ratio of 0.82. The moratorium on "all" ACR's refers to all converter reactors with conversion ratios above those attained by the current LWR. See Appendix B for a more detailed discussion.

construction of nuclear reactors of all types. In all these cases, reactors currently under construction or in operation are expected to be completed and to operate over their normal economic lives. We will occasionally write of a "nuclear moratorium," in quotes, as a reminder of the latter stipulation.

Similarly, the environmental damages or risks caused by technologies based on coal and shale, in extraction, conversion, or combustion, may be curtailed by limiting the annual production of coal and shale (which in 1975 were 15.2 and 0 Quads, respectively). Again, the MRG did not try to determine at what level of coal production and consumption the various risks and damages outweighed the economic benefits of further additions to coal use. Neither did it examine the feasibility of the high growth rates of coal production projected in several of its scenarios. To prepare the ground for possible later use of studies of this kind, the MRG defined, somewhat arbitrarily, two "limited coal and shale production" scenarios as two possible responses society may have to the risks associated with these technologies.

The first (2a) restricts the increase over time in annual production of coal to a figure of 40 Quads of primary energy equivalents in 2000, and ultimately to 50 Quads, and limits shale production to 2 Quads of primary energy equivalent in 2000, and ultimately to 12 Quads. The "limits" therefore restrict the future growth of coal to reach production levels that are no greater than three to four times current production. Production limits in the interim periods are defined by a mathematical function* that fits these two constraints.† This "medium production limit" is the one simulated by the largest number of models, and, unless otherwise specified, the "medium" limit is the one we refer to simply as "coal and shale limits."

A second, much more restrictive, coal and shale oil scenario (2b) was defined by the MRG at a late stage in the study. Designed to simulate energy price responses and interfuel substitution under conditions where strict limitations on the use of coal and shale oil are in place, this scenario limits the annual production of the two fuels combined to no more than 25 Quads. In Chapter III, we treat the results of this scenario as a mechanism for testing the sensitivity of the projections to the degree of restrictiveness embodied in a policy limiting coal and shale oil production.

Policy no. 3 envisions a combined environmental response to nuclear, coal, and shale oil technologies by placing the above restrictions on all three simultaneously. Policy no. 4 represents just the opposite, a relaxation of present environmental safety standards applicable to coal and nuclear technologies. Specification of this policy option responded to a request from the Synthesis Panel to simulate a scenario in which *private* costs to energy producers and consumers are reduced by specific policy actions other than the imposition of price controls.

*Technically known as a "logistic" curve.

†SRI implemented this constraint by altering the coal and shale supply curves.

That such environmental policies would produce large *social* costs not adequately reflected in market prices is well recognized by the MRG, but the information currently available to MRG does not allow us to make satisfactory estimates of these social, nonmarket costs in terms of dollar equivalents. This major deficiency in the current state of the art of energy modeling was already discussed in Chapter I.

Finally, policy no. 5 represents government actions that induce conservation over and above the level of voluntary conservation that results from higher energy prices. Thus, when consumers voluntarily insulate their homes after increases in fuel oil prices, they may do so because higher energy prices justify some capital expenditure to restrict their consumption of energy in this fashion. But when the government legislates energy efficiency standards for automobiles and appliances or effectively limits highway speeds by force of law, the resulting reduction of energy consumption is non-price-induced. The MRG also includes under the rubric of "non-price-induced conservation" the effects of government-sponsored educational efforts to provide consumers with information on various conservation measures and to inculcate a conservation ethic among energy consumers.

To represent quantitatively the policy of non-price-induced conservation, the MRG specified that such a policy would reduce aggregate energy consumption in 2010 by 10 percent below the level that otherwise would have prevailed in response to the prices projected for that year. In the years between 1976 and 2010, this conservation effect is presumed to be phased in gradually and linearly, reaching its maximum of 10 percent in energy savings by 2010.

Further description and definition of the policies and policy variables can be found in Appendix B.

b. Realization Variables

Listed in Table II.1 below are the "pure" realization variables and their "Base-Case" and "alternative" values. Except where indicated otherwise, all "Base-Case" values represent reference values agreed upon in each case by a majority of the MRG members, in the light of information received from other panels of the CONAES study and other sources. "Alternative" values are labeled "low" or "high." These values have been used in Chapter III for economic evaluation of technological options without assigning percentile values to them.

For the analysis of R&D decisions in Chapter IV, the R&DSG has attached judgmentally determined probabilities to all of the values--low, Base Case, and high--for the most important uncertain realization variables. These probabilities are reported in Chapter IV. Blank entries under the "high" and "low" headings in Table II.1 refer to cases where the MRG did not consider values other than those specified for the Base Case.

For the values of most of the "technological" realization variables, the MRG obtained the approval of a representative from the panel most directly able to provide expert judgment. In the case of "nontechnological" realization or blend variables, some other sources were used.

TABLE II.1 Base-Case and Alternative Values[a] of Important Realization Variables (All dollar costs are in 1975 dollars)

	Low	Base Case	High
1. *Growth rate of real GNP, Percent per annum* (low and high values represent the 10th and 90th percentiles, respectively):			
1975-1980	3.7	5.70[b]	7.0
1980-1990	2.7	3.28	4.8
1990-2000	1.2	2.84	4.0
2000-2010	0.5	2.48	4.0
2010 and beyond	0.2	1.80	3.5
2. *Capital Costs of Electricity Generating Technologies* ($\$/kW^{e}$, in year 2000; all capital costs reported at standardized 65 percent capacity factors):			
a. Coal-fired plant, including equipment for sulfur removal, at 36 percent efficiency	--	520	--
b. Light Water Reactor (LWR)	--	650	--
c. Representative Advanced Converter Reactor (ACR), 0.82 conversion ratio	--	715	--
d. Fast Breeder Reactor (FBR)	--	810	--
e. Solar-produced electricity	795	1,730	3,700
3. *Differential technical change in the energy sector (percent per annum) over that in the rest of the economy*	--	0	--
4. *U.S. oil and gas resources (in units of 10^{15} BTU) extractable at a cost up to \$2 per 10^6 BTU)*		1,720	
5. *U.S. uranium resources, in 10^6 tons of U_3O_8, recoverable at:*[c]			

Forward Cost of (\$/lb):	Minimum Acceptable Price of (\$/lb):	Low	Base Case	High
10.00	15.00	--	1.28	--
15.00	22.50	--	2.05	--
30.00	45.00	1.60	3.70	6.50[d]
50.00	75.00	--	5.46	--
80.00	120.00	--	7.80	--
100.00	150.00	2.10	9.24	8.50
150.00	225.00	2.25	12.57	9.10

TABLE II.1 Continued

	Low	Base Case	High
6. *Extraction and delivery costs of coal ($/10^6 BTU), 1985 and beyond:*			
a. *Low-sulfur coal,* for utilities (national average)	--	1.13	--
b. *High-sulfur coal,* for utilities (national average)	--	0.75	--
7. *Total average cost of synthetic fuels, delivered* ($/10^6 BTU; "availability" defined as production capacity of 0.5 × 10^15 BTU/yr):			
a. Coal gasification, available in 1990	--	3.70	--
b. Shale oil or coal liquefaction, available in 1995	--	3.70	--
8. *Price and income elasticities of demand*	No uniformity achieved		
9. *Average cost of clean nonelectric Alternative Energy System* ($/10^6 BTU):	2.00	5.00	8.00

NOTE: For detailed descriptions of the assumptions underlying some of these figures, see Appendix B.

[a]Blank spaces in the "low" and "high" columns indicate that no low or high alternative value was specified by the MRG.

[b]The Base-Case GNP projection implies an average annual growth rate of 3.8 percent over the 1970-1980 period. A substantial part of the rapid growth projected over the 1975-1980 period is due to an assumed recovery from the deep 1975 recession rather than from rapid growth of potential GNP. For a detailed discussion of the assumptions underlying the GNP projections over the entire period, 1975-2010, see Appendix B.

[c]The "minimum acceptable price" at which uranium resources would be supplied is estimated to be 50 percent higher than the "forward cost" concept that is conventionally used in ERDA documents. It is likely, however, that in the upper price ranges, the ratio of minimum acceptable price to forward cost is much closer to 1.0 than to 1.5. This is due to the fact that, for high cost and largely unknown deposits, the forward cost (which is essentially the equivalent of the economist's concept of marginal cost) is likely to include costs of exploration and development, costs that are excluded in computing forward costs for *known* deposits sold at much lower prices. Not having a well-agreed-upon procedure for translating forward costs into minimum acceptable prices in the upper price ranges, however, the MRG reported all minimum acceptable prices at 1.50 times forward costs.

[d]The figures reported in this column represent the "high" case because of the large quantity of low-cost resources.

For example, economic, econometric, and modeling studies in the literature were used as source material for the macroeconomic variables.

Table II.1 states the values assigned to the "pure" realization variables and brief indications of their definitions. More detailed discussion of the variables follows. A full statement of definitions, sources, and thinking behind the estimates is given in Appendix B.

Each of these variables is now discussed in turn.

1. Growth Rate of Real Gross National Product (GNP)

The Gross National Product (GNP) measures, for any one year, the sum of (1) the total value of all finished consumption goods and services reaching individual or collective consumers, and (2) the total value of all capital goods purchased by the firm, governmental agency, or individual who invests to add to production capacity, or for use. In a time series for real GNP, roughly speaking, the quantities of these goods and services for successive years are multiplied by their prices or unit values in some recent base year, and these "values" are added for each year in the series.

The standing of this variable as a purported measure of aggregate welfare of the population has suffered over time, and some work has been done to define better measures of economic welfare,* by adding imputations for the services of the consumer's capital, for the personal value of leisure, and for the value added by work done in households.

For the present purpose of identifying a measure of aggregate economic activity that can serve as a principal driving variable for demand for energy, the GNP seems as good as any other measure of aggregate welfare. The principal question, which bears equally on all such measures, is whether the causation flows only one way, from GNP (say) to energy use, or also the other way, in the sense that direct interventions that curtail energy use would in turn have a negative effect on GNP, especially at low levels of energy availability. This question will be addressed from the conceptual point of view in Section III.2 and in a quantitative way in Section III.9. Meanwhile, to begin the process of scenario construction, the MRG has used GNP as an important determinant of energy demand and consumption, while as yet disregarding possible feedbacks from energy use to GNP.

Accordingly, GNP growth has been classified as a *realization* variable, assumed to be determined in the long run by a combination of three factors largely independent of energy policy decisions: (1) population growth, (2) technological change, and (3) labor-force participation rates. The projections of GNP growth set forth above were made by estimating potential GNP on the basis of future trends for each of these variables, assuming no more than a floor-level of unemployment of labor.

*See, for instance, W. Nordhaus and J. Tobin, "Is Growth Obsolete?," in *Economic Growth*, published for National Bureau of Economic Research, Columbia University Press, New York, 1972.

Of course, in the short run, aggregate demand is the dominant determinant of GNP growth rather than potential GNP. For this reason, all estimates of GNP growth provide for higher growth over 1975-1980 than in subsequent periods, to account for further recovery from the recession of the last few years. Recessions may, of course, recur in the future, but the MRG assumes GNP to be at its potential level from 1985 on, to avoid underestimating the investments needed to meet future energy demand.

2. Capital Costs of Electricity-Generating Technologies

Since electricity is expected to play an important role in the next 50 years in a wide range of energy futures, the future capital costs of alternative technologies for the generation of electricity become an important realization variable in the study. These technologies include coal-fired generation plants, current Light Water Reactor (LWR) nuclear plants, a representative Advanced Converter Reactor (ACR) with a conversion ratio of 0.82, Fast Breeder Reactors (FBR's), and solar-produced electricity.

A representative of the Supply Panel verified each of the central and alternative estimates of the capital costs of these technologies. All capital cost estimates (in 1975 dollar equivalents) include interest costs incurred during construction and are adjusted for differences in efficiency and capacity factors to facilitate cost comparisons.* In addition, the capital and/or operating costs of all nuclear technologies also include allowances for the costs of police protection, waste disposal, and design costs required to meet current NRC safety standards.

3. Differential Technological Change in the Energy Sector

Historically, technological change has proceeded at a substantially faster pace in the energy sector than in the rest of the economy.† This has principally been due to continuing reductions in costs achieved through the economies of scale in the development of, and improvements in, electricity generation. It also has been due to the presence and

*For example, the Supply Panel reported that the Base-Case capital cost of a coal-fired generating plant with sulfur-removal equipment, operating at 65 percent of its capacity, would be $520/kW^e$ in 1975 dollars. The Solar Resource Group reported that a Base-Case capital cost for a solar-electric technology operating at 45 percent of capacity would be $1,200/kW^e$ in 1975 dollars. To make these capital cost estimates comparable at a 65 percent capacity equivalent, one must multiply the solar-electric capital cost of $1,200/kW^e$ by 0.65/0.45, yielding a figure of $1,730/kW^e$, as reported in Table II.1.

†See John W. Kendrick, "Productivity Trends in the United States," pp. 166-167.

growth in importance of components of the GNP, such as government services, in which, by the accounting conventions for the measurement of "output" of these services, no productivity growth has occurred at all.

For purposes of the CONAES study, the MRG adopted the assumption that cost reductions in the energy sector will no longer outpace those in the rest of the economy. The differential in technical change between the energy sector and that in the rest of the economy is therefore assumed to be 0 in the period 1976 through 2010.*

4. U.S. Oil and Gas Resources

Domestic oil and gas resources play a more important role for some policy questions than for others. In particular, the rate and cost of extraction of these resources are of central importance to the assessment of the value of synthetic fuels technologies. They are also potentially important variables in evaluating the benefits of advanced nuclear reactors in scenarios where coal and shale oil production are limited in response to one or more policy concerns.

5. U.S. Uranium Resources

The specification of the uranium supply curve (matching cumulative uranium resources against uranium costs) proved to be one of the most important and yet difficult tasks of the MRG. Uranium availability and cost are important, because they critically affect the economic benefits of developing advanced nuclear reactors designed largely to economize on, if not augment, available uranium resources. Estimation of uranium availability and cost is quite difficult, however, and there are considerable uncertainties that have yet to be resolved.

In spite of these difficulties, the MRG arrived at "high," "central," and "low" values of uranium availability, recognizing the inherent uncertainties involved. The central value was identified by combining information about low-cost uranium deposits in the National Uranium Resource Evaluation (NURE) Preliminary Report with the judgment of those asked by or represented in the MRG. (A detailed description of the procedure used to arrive at a central estimate of uranium resources appears in Appendix B.) High and low values for uranium availability were also specified for purposes of testing the sensitivity of the results to the uranium-availability assumption.

The uranium resource estimates are presented in Table II.1 as functions of "minimum acceptable prices," which the MRG assumed to be 50 percent higher than "forward costs," the cost concept ordinarily used in ERDA documents.

*SRI, however, allows for some differential technical change in specific technologies. This built-in feature of the model was too difficult to change for the MRG simulations.

6. Extraction and Delivery Costs of Coal

Since an important policy question addressed by the CONAES study concerns the relative costs of coal- and nuclear-fueled electricity generation, it was important to agree on a common set of cost estimates for the extraction and delivery of raw coal. The MRG selected coal costs applicable to utility use as representative figures for coal costs in all coal uses (except synthetic fuels). Coal costs for synthetic fuel production are included in the aggregate cost figure for synthetic fuels (see "Average Total Cost of Synthetic Fuels" below).

The coal costs reported in Table II.1 are strictly costs of extraction and delivery, and they exclude terms for economic rents (or royalties) that may arise in certain resource-constrained scenarios and that are estimated by most of the models (see "Long-Run Energy Price Trends," Section III.7). Moreover, the cost estimates do not account for possibly rising environmental costs associated with higher rates or cumulative amounts of coal production. It would have been desirable to include such costs, but limited time and data prevented the MRG from implementing the concept.

7. Average Total Cost of Synthetic Fuels

Of potential importance in all of the scenarios are the costs and dates of availability of various technologies for producing synthetic fossil fuels. These figures, reported in Table II.1, were obtained directly from the Supply Panel.

The costs and dates of availability of synthetic fuels promise to have their greatest impact on the degree to which electricity, a more desirable energy carrier, will remain--or become--competitive for many purposes in the future. They will also affect the degree to which the fossil-fuel powered automobile will be used in the remainder of this century and into the next. Finally, they may also be important in assessing the value of various nuclear technologies if coal and shale oil production are limited in response to one or more policy concerns.

8. Price and Income Elasticities of Demand

Like most other goods energy is, in most of its forms, in greater demand when incomes rise. The economist's notion of income elasticity measures the strength of this demand response to income changes. A long-run income elasticity of 1 expresses that, if price remains constant, a 10 percent higher income is ultimately associated with approximately 10 percent higher demand for the good in question (although adjustment over time to an abrupt change in income will be subject to a time lag of up to 10 years). Income elasticities less than or greater than 1 then imply a proportional fraction or multiple of a 10 percent higher demand.

Similarly, a long-run price elasticity of demand for energy equal to -1 indicates that a 10 percent higher price leads, after a period of adjustment, approximately to 10 percent lower energy consumption, so as to leave expenditure on energy (price times quantity) unchanged. In general, a price elasticity of $-e$ leads to a demand lowered by $10e$ percent. Total expenditure on energy products, then, varies in proportion to $1 - e$, upward if e is less than unity and downward if e is larger than unity.

Since both price and income elasticities of demand are defined as ratios of percentage changes, their numerical values are independent of the units of measurement of quantities or prices. On double-logarithmic graph paper, the elasticities show up as slope coefficients.

Clearly, the magnitude of price and income elasticities should have an important impact on the modeling projections. Simple arithmetic, for example, will show that projected energy demand in 2010 will vary within a 20 percent range if estimated income elasticities are varied between 0.9 and 1.1. And in 2010, such a 20 percent variation can translate, in the absence of price effects, into an absolute variation of 30 Quads (if before variation the projection is 150 Quads). This is nearly *half* of the 1975 level of energy consumption in the United States. Similar examples could be given about variations in price elasticities, where the range of uncertainty is still larger.

Income and price elasticities of demand are not constants of nature, such as the speed of light or the period of the moon. They describe aggregate responses that will vary between individuals, over time, and between different regions or countries. Their estimation from statistical data with the methods of econometrics is therefore a never-ending task. The elasticity differences among the six models illustrate the unfinished character of the task even with regard to presently available data. Some general conclusions stand out, however. There is extensive empirical evidence that for nearly all fuels and consuming sectors, the long-run price elasticities are negative and not zero, and the income elasticities are by no means always unity. Valuable and more definite information can be obtained from inspection of available studies. Table II.2 presents a compilation of elasticity estimates for four classes of fuels and three consuming sectors made by a variety of investigators. The entries in the table are taken from a more exhaustive study of this subject conducted for CONAES by Lester Taylor.*

The table reports that investigators have consistently found a negative response of fuel demands to fuel prices and positive response of per capita fuel demands to per capita income. It also illustrates that these responses are significantly stronger in the long run than in the short run.

An informative feature of the table is that it reports the type of data used by each investigator. Elasticity estimates from three types of data are recorded: time series, cross section, and pooled cross section and time series. These data differences are potentially important, since the use of either time-series or cross-sectional data alone may not produce enough range in the values of the variables to permit meaningful estimates. The use of pooled data may remove some of these

*See Lester A. Taylor, "The Demand for Energy: A Survey of Price and Income Elasticities," prepared for CONAES, April 1976.

TABLE II.2 Selected Estimates of Short- and Long-Run Price and Income Elasticities for Different Fuels and Consuming Sectors

Type of Demand and Principal Investigator(s)	Price Elasticity		Income Elasticity		Type of Data	Source
	Short Run	Long Run	Short Run	Long Run		
Electricity						
Residential (average price)						
Uri	-0.61	-1.66	0.04	0.12	TS: Monthly, aggregate U.S.	"A Dynamic Analysis for Electric Energy by Class of Consumer," Working Paper 34, Bureau of Labor Statistics, January 1975
FEA	-0.19	-1.46	0.30	1.10	CS-TS: Census regions of U.S. annual	*1976 National Energy Outlook*, FEA, 1976
Residential (marginal price)						
Taylor *et al.*	-0.07	-0.78	0.10	1.18	TS: States, annual	Taylor, L. D., Blattenberger, G. R., and Verleger, P. K., Jr., "The Residential Demand for Energy, Report to EFRI, January 1977
Commercial						
Uri	-0.34	-0.85	0.79	1.98	TS: Monthly, aggregate	"A Dynamic Analysis. . ."
FEA	-0.24	-0.38	0.73	1.63	CS-TS: U.S. Census regions of U.S., annual	*1976 National Energy Outlook*

Industrial						
Uri	-0.35	-0.69	1.32	2.63	TS: Monthly, aggregate	"A Dynamic Analysis. . ."
FEA	-0.15	-1.03	1.00	1.00	CS-TS: Census regions of U.S., annual	*1976 National Energy Outlook*
Natural gas						
Residential						
Houthakker/ Taylor	0	0	0.33	3.11	TS: Aggregate U.S.	*Consumer Demand in the United States, 1970*
FEA	-0.16	-1.26	0.30	2.42	CS-TS: Census regions of U.S., annual	*1976 National Energy Outlook*
Commercial						
Felton	NE	-1.45	NE	NE	CS: States	"Competition in the Energy Market Between Gas and Electricity", *Nebraska Journal of Economics and Business*, Autumn 1965
FEA	-0.38	Large	0.73	Large	CS-TS: Census regions of U.S., annual	*1976 National Energy Outlook*
Industrial						
MacAvoy/ Noll	NE	-1.78	NE	0.68	CS: Pipeline sales contracts	"Relative Prices on Regulated Transactions of the Natural Gas Pipelines," *The Bell Journal of Economics and Management Science*, Spring 1973

TABLE II.2 Continued

Type of Demand and Principal Investigator(s)	Price Elasticity		Income Elasticity		Type of Data	Source
	Short Run	Long Run	Short Run	Long Run		
FEA	-0.17	-0.58	1.00	1.00	CS-TS: Census regions of U.S., annual	*1976 National Energy Outlook*
Coal and fuel oil						
Reddy (total coal)	-0.39	-0.91	0.53	0.60	TS: Aggregate U.S.	"The Demand for Coal in the U.S.: An Econometric Analysis," in *Proceedings of A. ME Council of Economics*, February 23-28, 1974
Transportation fuels						
Houthakker/ Kennedy	-0.47	-0.80	0.74	1.33	TS-CS: 12 OECD countries	"Demand for Energy as a Function of Price." *Energy Delta*, American Astronautical Society, 1975
Aggregate of all fuels						
Manufacturing						
Griffin/ Gregory	NE	-0.79	NE	NE	TS-CS: 9 industrialized countries	"An Intercountry Translog Model of Energy Substitution Responses," *American Economic Review*, December 1976

NOTE: All price elasticities in this table refer to the price at the point of end use.
LEGEND: CS = cross section, TS = time series, CS-TS = pooled cross section and time series, NE = not estimated.

problems and therefore may produce elasticity estimates that are surrounded by less uncertainty.*

Due to these differences in data sources, equation specifications, and periods of estimation, the sample elasticity estimates reported in Table II.2 exhibit, for the same fuel and sector, a considerable degree of variation. Some of this variation can be attributed to differences in data sources: on balance, cross-sectional data, particularly across countries, tend to yield higher price and lower income elasticity estimates than time-series data. In interpreting the data in Table II.2, it should be kept in mind that, for the categories 1 through 4 that represent elasticities of demand with respect to particular energy carriers, the price elasticity is the so-called "own price elasticity," describing response to change in the price for that category of energy, with the prices for all other categories imagined to be held constant. The table does not record "cross-price elasticities," expressing the response of demand for a given type of fuel to a price change in just one other type of fuel only. Normally, these cross elasticities are positive numbers, reflecting substitution of the "other" fuel for the given one.

Taken as a whole, however, the estimates presented in Table II.2 support and corroborate the use of the nonunit income and nonzero price elasticities built into the models used by the MRG. In particular, the price elasticities in the DRI, FEA, Nordhaus, and SRI models, reported in Appendix C, are well within the range of variation exhibited in the table. The same is also true for the income elasticities incorporated in these models. Since elasticities in ETA are defined in a highly aggregated fashion at the point of primary energy extraction and not at the point of end use (as are the elasticities reported in Table II.2), comparing the ETA elasticities with those in Table II.2 is a difficult task.† A first effort to make such comparisons, reported in Appendix C, suggests that ETA's elasticities imply fuel and sector elasticities lying at the lower range of estimates recorded in the table. Finally, as described in both Appendixes A and C, although DESOM does not permit energy services at the point of end use to be responsive to energy prices, the modeling of interfuel substitution means that demands for total energy as well as individual fuels do change if prices change. Because this substitution is discontinuous, however, it is not an easy matter to express these responses through the price elasticity concept.

In view of the importance of the elasticity assumptions, the MRG made every effort to standardize the price and income elasticities of demand incorporated in the models. However, given the ways in which the models differ in structure, it was not possible within the limits of time and cost to attain uniformity in these elasticities. Indeed, it was sometimes difficult to establish comparable definitions because of differences

*William Nordhaus utilized this feature of pooled energy data in estimating energy demands in the 12 OECD countries. See the reference in Appendix C, Section C.4.

†For a fuller description of ETA's elasticities, see Appendix C.

in demand sectors between models. Some models, for example, report demand disaggregated by functional end uses (e.g., vehicle miles driven, space heating demanded, etc.); others disaggregate demand by broad end-use sectors (residential, commercial, etc.) and/or by specific fuels.

Nevertheless, to illustrate how important the elasticity assumptions can be, we report in Section III.11 comparative results from simulations using the ETA model under two different sets of elasticity assumptions. This exercise is supplemented in Appendix C by a detailed discussion and comparison of the elasticities (reduced to common definitions) incorporated in the various models.

9. Average Cost of a Clean Nonelectric Alternative Energy System (AES)

There is obviously a high degree of uncertainty about which of all possible energy technologies may become feasible between now and 2010, and at what cost. At the same time there are a number of energy sources, whether already tapped or still waiting in the wings, whose existence adds up to an assurance of the availability of energy in substantial quantities, at sufficiently higher energy prices.

To take account of these considerations, the MRG defined another conceptual realization variable, namely, the average cost of a clean nonelectric Alternative Energy System (AES). This variable is treated as an alternative source of supply and represents a variety of energy technologies of potentially unlimited size in the aggregate. One or more of these would enter the marketplace if there is a demand, at an equivalent cost, in the Base Case, of $\$5/10^6$ BTU, which is high relative to the options considered so far. By way of examples, the AES can include, among other items, solar heating and cooling, secondary or tertiary oil and gas recovery, possibly still new oil or gas discoveries, and also wind, geothermal, biomass, cogeneration, or any other new technology or supply source not yet implemented, designed, or even discovered. With some algebraic license we can include under the AES concept non-price-induced conservation in excess of the 10 percent already specified, a subtraction from demand being equivalent to an addition to supply.

All of the models except SRI and DESOM utilize the AES concept. In the SRI and DESOM models, the alternative energy systems are defined explicitly, and individual operating parameters and cost estimates are incorporated for these technologies. None of the other models, however, has been asked to specify a particular one of the AES alternatives as a likely least-cost option in any of the scenarios. Rather, the models have been fed only the judgmental estimate that at least one clean nonelectric energy option is available at any time after the year 2000 and that the least-cost collection of these options needed for any particular scenario will average out at the stated cost level in the Base Case, with lower or higher alternative possible cost levels used for sensitivity testing. The use of a single cost level for each scenario rather than a supply schedule is a simplification justified by the lack of operating experience with regard to the options in question.

One important role of the AES variable as used by the MRG is that it places an effective ceiling on energy prices in all of the projections.

In particular, that ceiling is the average cost figure that is assumed in defining the AES. For models used by the MRG, this avoids the possibility that, in some highly energy-constrained scenarios, energy prices would tend toward unbelievable levels because some types of plausible ceilings were not in place.* Thus, the device of the nonelectric AES is both an explicit admission of our uncertainties and an attempt to incorporate those uncertainties into the models at a level of aggregation that is not misleadingly detailed.

The AES is a summary statistic describing the adequacy of conventional sources of supply increase or demand reduction in achieving an energy system balance. If the AES enters a model's solution as a large component of total consumption, this indicates a situation where the energy sources included in the model are insufficient and new sources must be found or extraordinary demand reductions must be introduced. The AES concept does not dispose of such scenarios, but it does help us identify them within the limitations of our models.

c. Blends of Policy and Realization Variables

Several important driving variables could not easily be classified as either "pure" policy or "pure" realization variables, since they contained elements common to both of these categories. For that reason, the "Base-Case" values of these variables, listed in Table II.3, do not strictly represent explicit "central" value estimates. Rather, they reflect a "blend" of estimation and judgment, which produces a set of values thought to be of interest for evaluating the estimates of impacts of a variety of policies. "Alternative" values were also specified for sensitivity tests to examine how these policy impact estimates changed when the values of these "blend" variables were altered.

*This difficulty arises in these scenarios because of the way the demand functions are specified in the models. Specifically, the models (except for DESOM, which sets end-use price elasticities at 0) assume constant price elasticities, generally below 1 in absolute value. The lower these absolute values are, the more even moderate cutbacks in supply can force large increases in prices. In lieu of arbitrarily changing the form of these demand functions in the high-price regions where we have no historical experience to fall back on, one reason the MRG settled on the AES variable was to prevent prices from rising to levels where the price elasticities deemed appropriate in the realistic price range would force an unrealistic proportion of income to be spent on energy alone. By specifying a central value of $5 per million BTU as the average total cost of the clean nonelectric AES, the MRG is essentially preventing energy prices in these central value scenarios from rising over the period 1975-2010 to more than 2½ times their 1975 levels.

TABLE II.3 Base-Case and Alternative Values of Important Blend Variables

	Base Case	Alternative(s)	
1. *Discount rates* (percent per annum); two alternative values are presented:			
a. Pretax (for use in projecting pricing and investment decisions in the energy sector)	13	13(a)	6(b)
b. Posttax (for discounting benefits and costs of alternative technology options)	6	13(a)	6(b)
2. *Quantity ceilings and prices of imported oil and gas:*			
a. Annual import limit, as a percentage of annual domestic natural and synthetic oil and gas consumption	33	100	
b. Cumulative import limit, 10^{15} BTU	1,000	6,000	
c. Price of oil, in 1975 dollars per bbl, inclusive of tariff	13	5	
3. *Year of commercial availability, at cumulative installed capacity of 6 GW^e*			
Representative Advanced Converter Reactor (ACR)	2,000		
Fast Breeder Reactor (FBR)	2,000		
Central station solar (at central value capital costs)	2,000		
4. *Ceiling on the growth rate in capacity of new technologies* (percent per annum), after commercialization (cumulative installed capacity of 6 GW^e, or its equivalent)			
First 5 years after initial availability	40		
Second 5 years after initial availability	30		
Third 5 years after initial availability	20		
Beyond 15 years after initial availability	10		

1. Discount Rates

Taking each of these "blend" variables in turn, consider first the two discount rates presented in Table II.3. The real pretax discount rate, most familiar to those in the private sector charged with making pricing and investment decisions, is taken to be the rate producers use to discount profits from investment projects. This is the rate that is equivalent to the cost of capital in the private sector.

The other discount rate, the posttax rate, is taken to be the rate at which consumers discount the benefits they expect to derive from their future consumption. Put another way, this rate reflects the relative desires of consumers to consume now or save for future consumption. A given consumption rate of discount implies that consumers would prefer to consume more in the present than they already do if they could not invest their wealth with that expected real rate of return. The MRG has implicitly assumed that the rate at which consumers discount their private consumption possibilities is identical to the rate they would use as social decision makers in discounting the social benefits from having alternative energy technologies.

The corporate income tax creates a disparity between the pretax return earned by private corporations and the aftertax return earned by stockholders of such corporations. If no revenue were collected in this way, the unfettered workings of the marketplace would (but for risk differentials) ensure the equality of the consumption rate of discount and the rate used by producers in making their investment decisions. In the presence of the income tax on corporate earnings, a wedge is driven between the pretax rate of return used in making investment decisions and the aftertax rate of return identified with the rate of discount on consumption.

The market distortions created by the corporate income tax required the MRG to agree upon a second-best solution to the problem of which discount rate should be used in the MRG simulations. For its central or Base Case, the MRG adopted the suggestions of Kenneth Arrow, who was invited to discuss the problem with the members of the MRG. In brief, Arrow suggested that both rates of discount be employed, but each in a different role. Since, R&D aside, most energy investments would be implemented in the private sector, Arrow suggested that the pretax rate of return be used to calculate capital costs of various technologies and, therefore, to determine which technologies would be used to meet assumed sets of energy demands. He further suggested that the aftertax rate of return be used to discount the streams of consumption benefits from adopting alternative energy technologies. This is an innovation in energy modeling and, more generally, benefit-cost methodology that attempts to face up to a problem that has previously puzzled workers in the field.

One possible objection to the Arrow proposal and the way in which it was implemented by the MRG is that it collapses the diverse number of discount rates that may be observed in society into two measures: the before and after tax rates of return. In an economy free of risks this is perfectly acceptable, because an efficient capital market will then, but for taxes, equate all discount rates. Realistically, however, differences in risk between industries can cause producers' pretax rates of return to diverge. Likewise, inadequate functioning of the consumer capital market can generate differences in observed posttax rates of discount. Despite these valid objections to the use of two discount rates, however, the MRG decided that to disaggregate further would be too complicated and would probably add little in further precision to the projections.*

Another possible objection to the Arrow suggestion is that it overlooks potential differences between the discount rates consumers apply in making nonenergy consumption and savings decisions and those they use in deciding between energy consumption and savings. This objection stems from the observation that the implicit rate used by consumers in deciding whether to invest in energy-conserving devices (insulation, solar heating, etc.) appears to be substantially above the nonenergy consumption rate of discount.

Here, too, the matter is one of market failures. The disparity between the two rates of discount is due predominantly to inadequate information about conservation possibilities and to liquidity constraints on consumers. In applying a single consumption rate of discount, the MRG implicitly assumed that such imperfections, present to some degree now, would decline in importance in the long run.

Given the "blended" nature of the discount rate variable used by the MRG, an alternative policy impact scenario, a sensitivity scenario, and a somewhat "blended" scenario were simulated. The policy alternative reduced the rate of return used for pricing and investment decisions from 13 percent to 6 percent in the energy sector, through the enactment, for example, of appropriate tax credits. Such a move could be motivated by a desire to stimulate private investment in the energy sector. A sensitivity scenario, on the other hand, was constructed by raising the discount rate on consumption to 13 percent, to test the sensitivity of the policy impact estimates to higher consumption rates of discount. Finally, a scenario was constructed in which the discount rate for pricing and investment decisions in the public utility sector alone was assumed to be 10 percent, with the 13 percent and 6 percent rates applying elsewhere as previously outlined for the Base Case. This was done to account for the fact that debt/equity ratios are higher in the utility sector than in the rest of the economy, and to reflect the prevailing rates of return allowed public utilities by state and federal regulatory agencies.†

*Alan Manne and Ed Cazalet took exception to this view in regard to the use of a separate discount rate for public utility investment and pricing decisions. Both suggested that due to the historically lower rates of return that have, due to regulation, prevailed in the electric utility industry relative to those in the economy as a whole, that a 10 percent rate of return be applied separately to that sector in all simulations. This suggestion was tried in several sensitivity scenarios but was not followed in the basic set of MRG scenarios reported in Section III.5.

†Since many of these discount rate variations were simulated before the MRG agreed upon a final set of central values, only a selective portion of the numerical results from these simulations are presented in Chapter III, "Findings." These are supplemented, at appropriate places throughout that chapter, by a discussion and evaluation of the qualitative results from these simulations.

2. Quantity Ceilings and Prices of Imported Oil and Gas

Assumptions related to oil imports (blend variable no. 2) may also have an important bearing on several policy issues. In particular, the prices and quantities of imported oil will have a significant effect on the decision to develop synthetic fuels technologies. Oil import parameters may also affect the assessment of alternative electricity-generating technologies, particularly when the use of one or more resources is tightly limited for policy reasons.

Both the price and quantities of imported oil and gas have policy and realization aspects to them. Although the price of imported oil is primarily set by the OPEC cartel, it is potentially subject to influence by the magnitude of import purchases from the United States. Similar considerations apply to the quantity of imported oil, which to a large extent can be directly influenced by U.S. government policies, but which is also subject to the forces of competition on the world market for limited quantities of oil.

The MRG specified Base-Case values for the price of imported oil and for both annual and cumulative import limits for oil and gas. As an alternative, the price was lowered and the annual and cumulative constraints were relaxed in order to test the sensitivity of the impact estimates to the possibility that the OPEC cartel may break down in the near future.

3. Year of Commercial Availability of New Electricity-Generating Technologies

The times at which the advanced technologies represented in Table II.3 become available (defined as the point when installed electrical capacity of an electricity generation technology reaches 6 GW^e, or an equivalent Quads-per-year figure for synthetic fuels) are crucial to the entire range of issues examined in the CONAES study. These dates, however, are subject both to policy influences and to uncertainties surrounding the rate of advance of successful scientific research and development. Base-Case values for each technology have been approved by a representative of the Supply Panel.

4. Ceiling on the Growth Rate in Capacity of New Technologies

Finally, a ceiling on the growth rate of new technologies (except the AES) was specified as an important driving variable for two reasons. First, including this "real-world" constraint in the list of driving variables allowed the MRG to use the associated "shadow price" (reported only in the optimization models) to estimate the value of the technology to which the constraint referred. Although it would have been desirable for other reasons to apply the growth rate constraint to the AES, the MRG used the AES concept primarily to limit foreseeable price increases, as well as to increase the menu of energy technologies.

Second, examination of the preliminary results of many of the early simulations revealed that the growth rate in capacity ceilings had an important effect on estimated policy impacts. In particular, most of the initial simulations were performed under the assumption that a flat 20 percent growth rate ceiling was applicable throughout the lifetime of the technology. Under this assumption, many of the results exhibited a "hump" in energy prices during the period 1995-2010, precisely the period when many of the new technologies are expected to be introduced. The "humps" were present because the 20 percent ceilings were binding in the early, but not later, years of the presence of the new technologies.

In later simulations, the growth rate constraint was relaxed in the early years for all new technologies. The figures shown in Table II.3 were suggested by the historical rate of introduction of the LWR. The MRG assumed that the LWR experience would essentially be duplicated by all of the new technologies considered in the study. As indicated in Chapter III, the somewhat more liberal growth rate constraints in the early years of new technologies "smoothed out" the price paths over the 1995-2010 period substantially.*

The growth rate ceiling is a blend of both policy and realization elements because, although the government can influence the rate of capacity expansion by tax policies, the rate of introduction of a new technology depends to a large extent on the size of the firm(s) that developed it, the number of firms in the industry, and the profit expectations of those in the industry at the time the technology is shown to be commercially viable.

II.3 STEP 2: POLICY SCENARIOS AND POLICY IMPACT ANALYSIS FOR GIVEN MODELS

Step 2 consists of policy impact analyses, as described in Chapter I. This step is implemented by putting the same policy question to each model that can accommodate it. The MRG expresses no opinions on the merits of the policies represented in these scenarios. Rather it makes use of the models available to it to estimate the impacts of pursuing any one of these courses of action. The differences in the answers obtained, in spite of the common values given to most of the driving variables in the models, are due to differences in structure of the models and to differences in the values of driving variables not on our list because these values could not be changed in the models without excessive cost.

From the policy, realization, and blend variables listed in the foregoing section, the MRG constructed a "Base Case" consisting of all the

*Due to the way in which the SRI model is constructed, no growth rate ceilings were incorporated in the SRI simulations. Instead, upward sloping supply curves for each technology were used to represent the economic limitations of introducing a new technology at a rate faster than it can be absorbed.

Base-Case values, and then a series of "policy" and "sensitivity" scenarios. The Base Case incorporates the following broad policy assumptions:

1. All nuclear, coal, and shale oil technologies allowed.
2. Maintenance of the environmental standards specified under current law.*
3. Presence of non-price-induced conservation efforts (as defined in Section II.2 above).

The various alternative policy scenarios consisted of different combinations of the policy variables with Base-Case values of the realization and blend variables. These scenarios were designed principally to characterize different "energy futures" under various types of restrictions, imposed largely for environmental reasons, on the construction and/or use of certain technologies. These scenarios are listed in Table II.4.

We repeat that the MRG expresses no opinions on the merits of the policies represented in these scenarios. Rather, it makes use of the models available to it to estimate the impacts of pursuing any one of these courses of action.

The first three policy scenarios so defined (nos. 1-3) allow comparisons of output and price paths of different technologies under different nuclear reactor configurations. These simulations provide insight into how the availability of nuclear technologies affects the rest of the energy sector and the rest of the economy. They also allow estimates to be made of the strictly economic benefit of having different types of nuclear technologies available for use.

The next three scenarios, nos. 4-6, repeat the exercise of nos. 1-3 except that limits on annual coal and shale oil production are imposed. These scenarios are designed to simulate how the energy sector reacts to restrictions of increasing stringency imposed alternatively or simultaneously on the use of three important energy technologies: nuclear fission, coal, and shale oil. The restrictions on coal and shale oil production in scenarios 4-6 refer to the "medium" production limit defined in Section II.2.† Two additional "policy sensitivity" scenarios

*These standards include current reactor safety standards for nuclear generating plants and both ambient and emission standards currently applicable to fossil-fuel technologies. Although some have questioned whether even current pollution control technology can achieve the present ambient standards, the MRG Base Case assumes that in the long run such difficulties can be overcome. Clearly, if these difficulties persist and if environmental concern does not abate, restrictions may have to be placed on coal use, an outcome that is incorporated in the "coal and shale oil limits" policy outlined in this section.

†This policy limits annual coal production to 40 Quads in 2000, ultimately to 50 Quads, and shale oil to 2 Quads in 2000, 12 Quads ultimately.

TABLE II.4 Definitions of Policy Scenarios

1. Base-Case policies--All technologies available
2. Moratorium on the FBR and ACR, Base Case
3. Moratorium on all new nuclear construction, Base Case otherwise
4. "Medium production" limits on the annual production of coal and shale oil ("low coal and shale limits"), Base Case otherwise
5. Coal and shale limits, moratorium on the FBR and ACR, Base Case otherwise
6. Coal and shale limits, moratorium on all new nuclear construction, Base Case otherwise

4'. Low production limits on the annual production of coal and shale oil ("low coal and shale limits"), Base Case otherwise

6'. Low coal and shale limits, moratorium on all new nuclear construction, Base Case otherwise

7. High GNP growth, Base Case otherwise
8. Low GNP growth, Base Case otherwise
9. Low GNP growth, coal and shale limits, moratorium on all nuclear technologies, and high-cost nonelectric AES

nos. 4' and 6', were defined by using the "low production" limit on coal and shale oil.*

Results of these and all other policy impact analyses made as part of step 2 are presented in both tabular and graphical form and discussed in Chapter III.

In addition to the strictly policy-oriented scenarios just described, the MRG also simulated three other scenarios whose principal purpose was to generate widely different projections in aggregate energy consumption. These are listed in Table II.4 as nos. 7-9. Scenarios 7 and 8 are designed to provide some idea of the range in future energy consumption and the mix of energy sources that can be expected from variations in the GNP growth rate alone. The last scenario is designed to trace out a

*In this policy, annual coal and shale oil production combined are limited to 25 Quads.

path with very low energy consumption; thus, it projects what energy consumption would be if not only GNP growth were low, but if a series of environmental restrictions were imposed and only high-cost alternatives were available.

Results from the simulations of scenarios 7-9 are also reported in Chapter III.

II.4 STEP 3: SENSITIVITY ANALYSIS FOR GIVEN MODELS

Given the novel nature of the work of the MRG, there was a period of considerable experimentation before the MRG settled upon a final set of scenarios and central and alternative values for all of the driving variables. Thus, before agreement was reached, a substantial number of simulations based upon somewhat different input assumptions were made by various models. Some of these simulations correspond exactly to those reported under the description of step 2, except for the data differences.* If these differences are lumped together, this set of earlier simulations can be viewed as a "sensitivity" test of the final set of simulations with respect to that aggregate of data differences. The drawback to this procedure is that it is difficult to determine exactly which of the many data changes among the aggregate of data differences is responsible for a change in the simulation outputs.

Nevertheless, while it is true that the data inputs for all scenarios differed between the earlier and final set of simulations, the differences are not so great that they should vitiate the sensitivity comparisons between scenarios performed at an earlier stage of the work of the MRG. Thus, the impact of low uranium availability in the Base-Case results in the earlier set of simulations should be roughly the same in the final set. Similar reasoning should apply to changes in the discount rate.

The earlier sensitivity work, then, can be looked on as having been a helpful guide to the MRG. Through it, the group was able to identify certain critical variables and, just as important, critical ranges of variable values. For that reason, as well as the time and resource constraints, the MRG did not redo all of its earlier sensitivity work after it had agreed on a final set of scenarios and driving variable values. But, where it is possible and appropriate, the qualitative results of the earlier phase of the MRG's sensitivity analysis are highlighted in the findings reported in Chapter III.

The relevant sensitivity scenarios and the corresponding policy scenarios tested in the earlier phase of the study are listed in Table II.5.

*These differences are primarily in (1) the uranium supply curve, (2) the specification of the ceiling on the growth rate of new technologies, and (3) the specification of a clean nonelectric AES.

TABLE II.5 Sensitivity Tests for Given Models[a]

Sensitivity Scenario	Policy Scenario Tested
1. Low GNP growth rate	Base Case
2. High GNP growth rate	Base Case
3. Low uranium availability	(a) Base Case (b) Medium production limits on coal and shale ("coal and shale limits"), Base Case otherwise (c) Low production limits on coal and shale oil, Base Case otherwise (d) Coal and shale limits, moratorium on all advanced reactors, Base Case otherwise
4. Low capital cost of the FBR	Base Case
5. Thirteen percent discount rate for both consumption and investment decisions	Base Case
6. Ten percent discount rate for pricing and investment decisions in the public utility sector only	Base Case
7. High costs of synthetic fuels	Base Case
8. Low capital cost of solar-produced electricity	(a) Base Case (b) Coal and shale limits, moratorium on all nuclear technologies, Base Case otherwise

[a]Note that the Base Case specified in Table II.5 refers to the Base Case of earlier simulations, and differs in some respects from the Base Case finally used.

II.5 STEP 4: SENSITIVITY ANALYSIS ACROSS MODELS

Step 4 involves the comparison of the answers given to the same policy impact question by different models. Because of the nonuniform model structures and data inputs for driving variables not listed above, this comparison is in effect a sensitivity analysis of the policy impact estimates with respect to differences in model structures and in certain data inputs. An important example is the case of income and price elasticities of demand for energy or its components. In Section III.11 we describe the possible inferences that may be drawn as to the significance of these parameters in assessing policy impacts. In Appendix C we record the explicit or implicit differences in some of the most important values of the elasticities incorporated in the models used by the MRG, and in Appendix D we outline in greater detail the procedure used in Section III.11 to isolate what factors were responsible for differences in model results.

Inputs to sensitivity analyses of this kind can be obtained by making horizontal comparisons in all tables in Chapter III in which policy analyses are made for more than one model.

Another important comparative purpose is served by having models designed for intermediate-term as well as for long-term comparisons represented in the MRG. By comparing the key output results of the intermediate-term models with those of the longer-run models for the period 1975-1990, the MRG was able to "validate," in a way, at least a segment of the projections made by the long-run models.

III FINDINGS

III.1 OVERVIEW

We begin (Section III.1) with a broad sketch that relates the principal conclusions to the principal assumptions; next (Section III.2) we discuss in an exploratory, conceptual way the important question whether or when GNP can be treated, as we do, as a given driving variable that is itself independent of energy consumption; thereafter (Sections III.3 through III.11) we present a sequence of more detailed tables and charts containing principal findings, accompanied by explanations of their salient features. In particular, in Section III.8 we present estimated economic evaluations of important technologies. In Section III.9 we return to the problem of possible feedback from energy use to GNP to give some quantitative estimates.

Table III.1 summarizes the principal findings in telegram style, and may serve as a somewhat selective table of contents for the present "overview." The first entry reports that, in all projections made with the Base-Case assumption of an average annual growth rate in GNP of 3.2 percent over the period 1975-2010, there is substantial continued growth in energy use, at a lesser rate. The average growth rates in total energy consumption over the same period vary between 1.7 percent and 2.9 percent per annum, depending on the model used and on the extent to which environmental or other policy constraints are imposed on the various supply technologies. Substantially lower growth rates of energy consumption are projected only in the low GNP growth case, with an assumed average growth rate of GNP of 1.8 percent, and this reduction occurs regardless of the presence or strength of the supply constraints we have considered.

Most of the growth rates in energy consumption that are projected under the Base-Case GNP growth rate for various supply alternatives are less than the 1950-1975 growth rate in energy consumption of 3.0 percent, and all of them are significantly less than the Base-Case GNP growth rate. In particular, the growth rates of electricity consumption between 1975 and 2010, projected under the Base-Case assumptions for *all* realizations, vary between models from 2.1 percent to 3.9 percent. Even though the higher growth rate projections were made by models that allow only a low degree of price responsiveness ("price elasticity of demand") by end-use consumers, they are still well below the 1975-1990 projections of 5.3 percent to 5.8 percent current in the electric utility industry.

TABLE III.1 Summary of Projections Based on MRG Assumptions

Projections		Assumptions (Base Case, or departures from Base Case, for definitions see Chapter II and Appendix B)
Growth in Energy Consumption		GNP grows
ranges between 1.7 percent and 2.9 percent per annum	*if*	at Base-Case rate of 3.2 percent per annum
is higher (lower)	*if*	faster (slower)
is higher	*if*	energy price increases are held down by policy
Energy Price Increases		
range between 0.5 percent and 1.0 percent per annum	*in*	Base Case
are nearer high (low) end of range	*as*	the price elasticity of demand for energy is lower (higher) in absolute value
are higher	*if*	one or more energy technologies are constrained for policy reasons
are higher	*if*	uranium resources turn out to be very limited
are higher	*if*	a moderate-cost Alternative Energy System cannot be found by 2000
Growth in Electricity Consumption		
ranges between 2.1 percent and 3.9 percent per annum, which falls short of the 6.2 percent rate over 1946-1975	*in*	Base Case
because		
Electricity Price Increase		
ranges between 0.9 percent and 1.2 percent per annum, in contrast with the 0.5 percent rate of *decrease* for 1946-1975	*in*	Base Case
GNP Growth		
is, *in the long run,* relatively independent of growth in energy consumption	*if*	the price elasticity of demand for energy is -(0.5 or greater)

TABLE III.1 Continued

Projections		Assumptions (Base Case, or departures from Base Case, for definitions see Chapter II and Appendix B)
The Energy Supply Mix is projected to		
depend strongly on coal	*in*	Base Case
involve considerable growth of nuclear power	*in*	Base Case
turn away from oil and gas	*in*	Base Case
eventually turn toward solar and possibly other plentiful sources	*in*	Base Case
Technology Evaluations:		
The net economic *benefit* from Advanced Nuclear Reactors is low	*unless*	GNP growth is very high
	and/or	uranium availability is low
	and/or	the price elasticity of demand for energy is low
The net economic *cost* of		
(1) a moratorium on new construction of all nuclear reactors is substantial		by itself
(2) the MRG "medium" limit on coal and shale oil production is substantial		by itself
(3) (1) *and* (2) together average about *twice* the sum of these substantial costs		applied simultaneously

NOTE: Environmental costs or benefits are not included or evaluated in these projections.

The projected energy growth rates imply a gradual long-run decline in the ratio of energy use to GNP in the future. Given a Base-Case GNP growth rate of 3.2 percent over the period 1975-2010, the projections of total energy consumption imply that by 2010 total energy consumption per unit of GNP is projected to be 5 to 40 percent lower than it was in 1975. This would continue the historical trend that has seen energy consumption per unit of GNP fall by approximately 15 percent between 1947 and 1975.

These findings differ from widely held views. The fact that we project lower growth rates in energy consumption than generally expected is only in small part due to the MRG assumption of a non-price-induced conservation effect, amounting to 10 percent of energy use by 2010. The principal reason for the difference is that (apart from DESOM) all models assume that a higher aggregate energy price leads in the long run to a lower demand for aggregate energy--an assumption supported by empirical studies in the field (see the discussion of elasticities in Section II.1, realization variable no. 8). A second reason is that most models also assume that a higher consumers' income leads to a less than proportionally higher energy consumption, the ratio of corresponding percentage changes ranging between models from 0.75 to 1.0.

Clearly then, the energy projections depend strongly on projections of income growth and energy prices. GNP projections were made, as described in Chapter II, by independent analysis of expected future trends in the variables that determine GNP: labor-force growth and technological change. As a working hypothesis, the members of the MRG took the view that, in a first approximation, neither of these two components of GNP appeared to be directly affected in the long run by developments within the energy sector. With respect to the energy sector specifically, therefore, the growth rate of GNP is treated as a driving rather than a driven variable. Further discussion and testing of this important hypothesis in Sections III.2 and III.8 tends to confirm it, provided the constraints imposed on energy supplies act gradually rather than abruptly (as they did in 1973), and, in the case of substantial cutbacks in energy consumption, provided that the long-run price elasticity of the demand for aggregate energy use is not materially below 0.5 in absolute value.

In contrast to the projections of GNP growth, projections of energy prices are interrelated with those of energy consumption, and are therefore made by the models themselves. The extent of the projected price increases in turn depends on the extent to which supply is constrained for reasons of environmental protection, energy independence, or other policy considerations. In the Base Case, energy prices as a whole are projected to rise at rates between 0.5 percent and 1.0 percent per annum. Electricity prices, in particular, are projected to rise at rates between 0.9 percent and 1.2 percent, in contrast with the 0.5 percent annual rate of decrease between 1946 and 1975. An upper bound on the price rise in any scenario and model is provided by the assumption that a number of higher-cost energy supply options not now in use constitute a practically unlimited AES from which one or more processes can be brought in if energy prices rise sufficiently high.

How close we will get to that "upper bound" will depend very much on two factors: (1) the responsiveness of energy demand to price changes (price elasticity), and (2) the strength of supply constraints. The long-run price elasticities incorporated in the models determine how much demand reduction can be expected over time as energy prices rise. If the demand response is large, resources will be used at a slower rate, and the need for an Alternative Energy System will be postponed. The converse is true if, even over an extended period, consumers do not materially change their levels of energy use as prices rise.

As Section II.1 indicates, the MRG was unable to standardize the long-run elasticities across models. A large part of the differences in model results can be explained by this lack of elasticity uniformity (see Section III.11). That these differences exist, however, is an important reminder of the need to strengthen the empirical basis of the elasticity estimates so as to narrow the range of these discrepancies.

On the supply side, perhaps the outstanding constraint facing us now is a limited domestic supply of petroleum and natural gas resources. In the MRG simulations, it is supposed that the remaining domestic and imported resources taken together amount to no more than $3{,}000 \times 10^{15}$ BTU (3,000 Quads, for short), all available at a cost equal to the 1975 import price of approximately $2 per million BTU. This amount includes not only today's proven reserves of approximately 400 Quads, but also the more speculative categories of "undiscovered recoverable resources" in Alaska and the outer continental shelf.

To effect a transition toward an energy economy based on non-oil-and-gas supply sources, there are many options available--each with its pros and cons. Among non-oil-and-gas sources the projected supply mix varies depending on what is assumed with respect to relative costs--including the costs of meeting future environmental and safety standards with the best-available technologies.

Our Base Case is one which leads to heavy reliance upon coal--both for electricity generation and for the production of synthetic fuels. Although this projection presupposes effective pollution abatement equipment, coal nevertheless appears on strictly economic grounds to be a relatively low-cost future supply alternative to oil and gas.

However, there must be limits to the rate at which coal production and use can be expanded. Moreover, a scenario in which coal predominates is likely to produce a variety of environmental problems in addition to those that can be overcome by pollution abatement techniques. In the "coal-and-shale-limits" scenarios, coal is restricted to the extent of not exceeding three to four times the current level of production. At this level there may yet be significant adverse effects from coal use. On the other hand, if such constraints are *not* imposed, coal production comes out substantially higher in most of the models. Accordingly, in most simulations, the introduction of coal and shale limits causes substantial increases in the projected prices of nonelectric energy and a moderate increase in the projected price of electricity. However, if new technologies become available at moderate costs (whether advanced converters, breeder reactors, fusion, direct or indirect solar, or others) in the period after 2000, electricity costs could then be stabilized. A large-scale substitution of electricity for nonelectric energy would then be likely to follow during the period between 2010 and 2030.

Until 2010, however, electricity growth is projected to proceed at a substantially slower pace than the approximate 7 percent per annum increase in demand experienced between 1946 and 1975. The principal reason for this finding is the projection of higher electricity costs and prices in the future in contrast with the historical decline in electricity prices prior to 1975.

In our Base Case, for example, both nuclear fission and coal are capable of providing electricity at approximately the current cost, expressed in 1975 dollars. Therefore, if either substantial limits are placed on coal production or a moratorium is imposed on further new construction of nuclear fission plants, electricity prices in the long run are not likely to rise much above current levels. However, if *both* coal restrictions and a "nuclear moratorium" are imposed, prices are projected to rise to the price ceiling provided by the alternative nonelectric energy system.

Clearly, then, the price at which a clean nonelectric Alternative Energy System (AES) will be available in 2000 plays a crucial role in the price projections and technology evaluations. Specifically, if instead of the Base-Case AES cost of $\$5/10^6$ BTU, the AES actually costs $\$8/10^6$ BTU in 2000, a simultaneous limitation of coal production and a "nuclear moratorium" could force energy prices to rise as high as the AES ceiling price of \$8 instead of $\$5/10^6$ BTU. Higher AES costs than assumed in the Base Case would also increase the value of advanced technologies, such as advanced nuclear reactors, in most scenarios. For most scenarios, however, the resource availability and price elasticity assumptions do not permit energy prices to rise even as high as $\$5/10^6$ BTU in 2010, giving rise to little demand for an AES technology.

Nevertheless, the need for precision about an AES remains, for in the event that society deems it necessary to place restrictions on coal and nuclear energy, the need for a clean nonelectric AES at moderate cost will become apparent. As Chapter II outlines, the models defined the AES in various ways: direct solar heating and cooling, non-price-induced conservation, new oil discoveries, and other as yet undefined technologies. Admittedly, the concept could have been defined with greater precision. That it was not is an indication of the large uncertainty we all face in attempting to project a menu of technologies 25 to 35 years hence. A low- or moderate-cost AES may not exist. Modeling is nevertheless a well-structured way of projecting what may happen under a variety of assumptions about the possible cost of an AES in 2000.

Other constraints and/or demand realizations can also impel electricity prices upward. Limited uranium availability, for example, or higher GNP growth than projected in the Base Case, can force electricity prices to rise unless another low-cost energy source becomes available. In particular, the principal relatively low-cost alternatives currently under discussion make up the package of advanced nuclear reactors, consisting of advanced converters and breeder reactors. To the extent that R&D on solar, fusion, or other sources were to prove successful, however, any of these sources could in due course provide low-cost energy instead of the advanced nuclear technologies. On the other hand, if solar-produced electricity remains a high-cost source of electricity, restrictions on nuclear and coal would lead to the introduction of solar, but only at costs that are substantially above the current costs of electricity.

Under the Base-Case cost assumptions used by the MRG, either advanced converters or breeders are nevertheless still much less expensive than solar electric, fusion, or other long-term Alternative Energy Sources. As a result, if nuclear fission is to be relied upon to a large extent, the issue of the best mix of reactor types arises.

From the calculations performed by the MRG, the projections indicate that, without electric or nonelectric Alternative Energy Systems, there is likely to be economic justification for a reactor that utilizes its direct or ultimate source of energy, whether uranium or thorium, more efficiently than the Light Water Reactor presently in use does. Within this category, the economic comparison based on available cost estimates (Table III.11) slightly favors an advanced converter (type unspecified but conversion ratio = 0.82) over the liquid metal Fast Breeder Reactor. The difference may well be overshadowed by uncertainties in capital cost or differences in R&D cost, timing, and probability of success, and, of course, by the difficult problems of handling and disposing of hazardous nuclear materials and the hard-to-assess risks of their proliferation or diversion.

The use of both types of reactors (ACR and FBR) is projected to be much greater if uranium resources prove to be lower than the levels assumed in the Base Case. A similar result holds if the interest rate applied in the calculation of capital cost for energy investment and pricing decisions is closer to 10 percent than to the Base-Case rate of 13 percent.

Closely related to the question of which types of nuclear reactors are projected to be most economical in different scenarios is the estimation of what economic value those reactors, if put to use, are likely to have for society as a whole. In particular, it is of interest to know what value of goods and services, discounted over a long horizon (say, to 2050 or 2060), society stands to gain by introducing different nuclear technologies. In Section III.8 we report the results of simulations by DESOM, ETA, and Nordhaus designed to answer that question. In these evaluations, a social discount rate of 6 percent is applied to all future benefits minus costs.

Under Base-Case assumptions about the values of all the realization and policy variables, ETA and Nordhaus estimate the lump-sum discounted value of all advanced reactors taken together (including both the FBR and ACR), *before* accounting for R&D costs, to lie between $2 and $4 billion. If coal and shale are constrained, the ETA and Nordhaus estimated range of benefits rises to between $5 and $8 billion.*

The ETA and Nordhaus estimates are predicated on the assumption that energy demands will respond to price changes much as they have in the

*Both sets of estimates do not include the value of plutonium recycling alone, which is as high as $17 billion when coal and shale oil are constrained. See Section III.8 for further details.

past. If energy demands prove to be inelastic, however, the economic benefits of an advanced nuclear technology are projected to be much more substantial. Thus, the DESOM model, in which the demands for energy services do not respond to price changes, estimates the economic value of advanced reactors to be $43 billion in the Base Case and $98 billion if coal and shale oil production are constrained.

The same method of estimation just described can be extended to determine the strictly economic cost of placing restrictions on the use of nuclear power generally or on coal and shale oil production. In these simulations, also reported in Section III.8, both ETA and Nordhaus between them project the costs of either a "nuclear moratorium" or limits on coal and shale oil to lie between $46 and $159 billion. These costs, translated into per capita (1976 population) terms, suggest lump-sum discounted costs of between $200 and $750. At a 6 percent rate of discount, this range of per capita lump-sum costs implies a range of annual losses in perpetuity of between $12 and $36, all in 1975 dollars. These estimated costs are considerably higher, however, if both a nuclear moratorium and coal and shale oil limits are in place. In that case, the ETA and Nordhaus lump-sum estimates climb to $358 and $457 billion, respectively, or on a per capita basis, to between $1,700 and $2,100.

The cost estimates made by DESOM are even higher than those of ETA and Nordhaus in the cases of nuclear moratorium and/or restrictions in coal and shale oil production. Again, this is because in DESOM demands for energy services do not respond to price changes. These estimates, reported in Section III.8, therefore provide a type of theoretical "upper bound" on the other estimates.

One shortcoming with the analysis in Section III.8, however, is that it covers only a selected subset of possible combinations of the values of realization variables. Since a more complete treatment of the issue of technology evaluation would seem to require a consideration of a larger number of cases, the MRG turned to a subgroup to provide a detailed decision analysis of both the FBR's and ACR's. Results of that work, presented in Chapter IV, confirm the broad results presented here and in both Section III.8 and Chapter IV do not incorporate estimates of adverse impacts on environmental and other policy objectives that may result from the use of the various energy technologies.

Finally, in reading the more detailed presentation of the energy projections that follow, it is important to keep in mind the accounting conventions used to report the results. Without elaborating at length here (a detailed summary of the conventions appears in Appendix E), all energy production figures contained in this report are reported in terms of Quads (10^{15} BTU) of "primary energy equivalents," the convention employed by the Bureau of Mines.

While this framework is widely used, it can produce results that, to the untrained eye, may appear highly anomalous or surprising. In particular, the conventions may allow different fuel mixes that can render the same bundle of energy services to add up to different total figures for energy production. A common example where this often occurs is where synthetic fuels are used instead of crude oil and natural gas. Since there are considerable efficiency losses in producing synthetic fuels from coal, the same level of energy services (e.g., passenger miles driven) would show much higher levels of consumption, measured by primary energy equivalents, when provided by synthetic fuels rather than by crude oil or natural gas. Similar problems arising with solar and nuclear power are discussed in greater detail in Section III.4.

As a result, much of the apparent difference between the aggregate energy production projections made by the models is due to the way in which energy investigators currently account for differences in fuel mixes. An attempt to determine how much of the difference in model projections can be explained in this way appears in Section III.11.

III.2 FEEDBACK FROM ENERGY USE TO GNP

In all MRG scenarios we have treated the Gross National Product (GNP) as a principal driving variable strongly influencing energy consumption but not in turn influenced by it. At the same time we have in various scenarios considered policies that reduce energy supply and hence consumption directly rather than via GNP. Specifically, we have estimated the effects of policies that curtail energy supply below what it would otherwise have been--curtailments imposed for reasons of environmental protection, energy independence, and other national policy objectives. In those cases, does not the diminished use of energy also reduce GNP below what it would otherwise have been?

Virtually everyone will agree that such a reverse effect exists. For our present purpose, however, the real question is how large or small it is, in various circumstances. One important distinction is whether the curtailment of supply is abrupt or gradual, and if abrupt whether it is foreseen or not. The OPEC embargo and related energy price rises in 1973-1974 were both abrupt and unforeseen. Their effect on the economy included temporary unemployment of fuel-specific labor, plant, and equipment and a reinforcement of the rise in the general price level. The unforeseen severity of the winter of 1977 added asperity and disruptiveness to the creeping natural gas scarcity, which, in a more general sense, could have been anticipated better and earlier than it was. Neither of these two events, however, provide evidence for the long-range effect of a gradual and foreseeable reduction of energy supply that reaches a substantial cutback in quantity over a period of, for example, 10 to 15 years.

The long-run effect of such gradual and foreseen curtailments in energy supply is a simpler phenomenon, if we make the assumption that over time the government will pursue and successfully implement fiscal and monetary policies that maintain reasonably stable and full employment of the labor force and the capital stock. Then even an actual contraction in energy use could still be absorbed without creating significant unemployment of either capital or labor, as long as it is gradual, and proceeds at a rate no higher than that implicit in the normal retirement cycle of the plant and equipment specialized for energy production and use, and in the normal rate of turnover of the attendant labor. In that case, the capital funds so released, and workers seeking alternative employment, could find reemployment in other production processes that serve increased demands for less energy-intensive goods and services.

However, to the extent that technical improvements in energy production or use are embodied in the type of capital stock used, such a contraction--or even just a slower growth--would slow down the implementation of advances in technique.

There are several routes of substitution that are followed concurrently in a gradual shift of capital, labor, and nonenergy primary resources away from energy-producing processes, both away from and toward some energy-using processes and toward other processes serving the same or alternative human needs or desires. A convenient baseline for their discussion is the path that would be followed by energy use per unit of real GNP if the real price of primary energy were to remain constant over time. Historically this imaginary path has moved mostly downward over the period 1922-1956, at an average rate of about 2 percent per annum,* and has stayed roughly constant for 1956-1970. As Schurr points out, this history reflects a period of technical advances of two kinds. One is the advances in energy efficiency in the production and direct use of energy in given forms. The other is the increase in general productivity (output from given inputs) in the economy as a whole, made possible in part by the gradual shift toward energy carriers with greater flexibility in use--such as liquid fuels and electricity--even though when looking at the full sequence of energy conversions the use of these superior fuels has a lower so-called energy efficiency associated with it.

As long as these trends happen to continue into the future in such a way as to keep the real price of primary energy roughly constant, the actual path of energy use will follow the baseline suggested above. If consideration of environmental protection and other policy goals suggest a curtailment of net energy growth below that of the baseline, the intrinsic (or "shadow") price of energy will increase--although actual prices may be made to lag behind for a time through regulation, price control, formal or informal rationing, possibly accompanied by interruptions of supply. The rationing, supply interruptions, or actual price rises of energy in its various forms will then set into motion a number of substitutions in the inputs to production per unit of output, in all stages of production from primary materials to finished capital goods, consumption goods, and services. By and large the substitutions will be away from energy inputs, substituting other materials (insulation), possibly capital (accelerating the trend toward greater energy efficiency where possible), and/or some labor.

Additional substitutions will occur in the consumption sphere. These are likely to be a mixture of price-induced, voluntary, and mandated substitutions that further diminish energy use. Slower driving of smaller, lighter, and more energy-efficient cars substitutes time and some inconvenience or loss of pride. More generally, there will be a shift away from goods that are energy-intensive in their production

*This rate is based on Edmonson's estimated price elasticity of -0.5 for U.S. demand for aggregate energy, an aggregate energy index computed by Nordhaus using 1963 value weights, and on a memorandum of Sam Schurr, first draft, January 21, 1977.

toward other goods of different design or materials that serve the same purposes. Finally, there will also be some substitution away from forms of consumption (indoor ice skating) served mostly by energy-intensive goods or processes to other forms of consumption (basketball) that before the energy price rise promised about equal satisfaction per dollar spent. The extent of these substitutions will be larger the higher the component of energy cost in the price of the good or service that is curtailed.

Many of these substitutions in production and consumption have been studied in greater detail in the report of the Demand/Conservation Panel. Two points must also be stressed here. First, the most important substitutions require time for the needed reallocation of capital to come about, and in some cases for the needed deregulation and/or new regulations to be carried through.

The second point concerns the aggregate effects of the substitutions between inputs per unit of output in most or all industrial processes and in many forms of consumption. To visualize the total shifts in interindustry and industry-to-final-consumer flows that result from these substitutions, we must multiply, for each type of input good or service, the percentage change in the input-to-output ratio with the percentage change in the output of the industry or process concerned and add overall processes and forms of consumption. Obviously, compared with the baseline path, total energy production and use will be less. Capital use per unit of output, or of energy use, will be higher, but total capital applied to the *production* of energy is likely to be less, although the *use* of energy might be larger than in the baseline path. But one would expect a net shift of capital and other resources from the processes of energy production and use taken together toward the meeting of human wants in other, less energy-intensive, ways.

At every next step in this sequence of plausible inferences one feels a stronger need for quantitative estimation of these aggregate effects. The critical parameter that describes the quantitative effect of all these energy-saving substitutions taken together, and thereby determines the feedback from energy use to GNP, turns out to be the long-run price elasticity of demand for energy. This elasticity incorporates price responses by producers as well as consumers. The producer's choice aspect is dominant in the choices between alternative processes of production that ultimately meet the same wants. (In this formulation the firm or person who builds residences and decides on heating and cooling systems must be counted as a producer.) The consumer's choice aspect dominates in the setting of the thermostat, in choosing between leisure time activities or residential locations, and in the response to prices of goods and services for which energy cost adds up to a noticeable component of the price.

In Section III.9 we report on a number of calculations, based on several models, that throw light on the quantitative importance of substitution in production and consumption taken together and thereby on the magnitude of the feedback from gradual curtailments of energy use to GNP or to total real income. The principal finding is that, for a price elasticity of demand for energy which, in absolute amount, is not less than one-half, the long-run decrease in future GNP due to gradually cutting energy use down to one-half of the baseline level is estimated to be, by 2010, of the order of 1 percent to 6 percent. Details about the calculations, as well as caveats in their interpretation, are given in Section III.9.

III.3 COMPARISON OF 1990 OUTPUTS BETWEEN INTERMEDIATE- AND LONG-RUN MODELS

Although the focus of the CONAES study is on what our energy future may look like in 2010, there is still an important role for projection techniques with shorter time horizons. The MRG was fortunate to have at its disposal two well-structured intermediate-term energy models, DRI and FEA, which trace out energy trajectories between now and 1990 in greater detail than some of the longer-run models. These two models were used to check the intermediate-term output results produced by the longer-run models.

Table III.2 compares the 1990 Base-Case output results for selected variables produced by FEA and DRI (both intermediate-term models) with the results reported by each of the longer-run models. Energy quantities and not prices are presented because of differences between the models in the way prices are defined. (See the discussion of energy prices, Section III.6.) The table reveals striking differences between

TABLE III.2 Comparison of Selected Outputs of Primary Energy and of Electric Energy for 1990, Base Case, Between the Intermediate- and Long-Run Models (Unit: Quads of Primary Energy Equivalents)

	Intermediate-Run Models		Long-Run Models			
	FEA	DRI	DESOM	ETA	Nordhaus	SRI
Domestic energy production	93.4	--	93.6	81.5	81.1	94.7
Coal production	25.3	25.2	20.5	20.1	28.1	29.8
Natural gas	22.2	--	31.6	46.6	21.1	19.3
Oil and natural gas liquids	29.4	11.6	24.9		5.2	24.0
Nuclear energy	12.3	--	11.3	10.1	26.7	16.7
Solar heating and cooling	0.0	--	0.1	0.0	0.0	0.8
Other	4.2	4.7	4.9	4.6	0.0	4.2
Oil and gas imports	21.1	--	18.9	23.3	15.1	14.2
Total energy consumption	114.5	107.9	112.5	104.8	96.2	108.9
Electricity generation (10^{12} kWh)	4.3	4.1	3.1	3.3	2.8	4.1

TABLE III.3 Total Consumption, Domestic Production, and Imports of Energy in 1975, 1990, and 2010, Under Different Supply Policies (Unit: Quads of Primary Energy Equivalents)

Policies: Base Case, Modified as Stated (Scenario Number Indicated at Left)	Actual	DESOM		ETA		Nordhaus		SRI	
	1975	1990	2010	1990	2010	1990	2010	1990	2010
1. Base Case									
Consumption	71.6	112.5	196.3	104.8	178.9	96.2	129.6	108.9	185.4
Production	58.8	93.6	171.2	81.5	167.0	81.1	106.7	94.7	166.6
Imports	12.8	18.9	25.1	23.3	11.9	15.1	22.9	14.2	18.8
3. Nuclear moratorium									
Consumption	71.6	112.0	193.2	103.9	175.4	90.1	125.6	--	168.3
Production	58.8	93.3	166.9	80.5	163.6	72.2	103.2	--	147.8
Imports	12.8	18.7	26.3	23.4	11.8	17.9	22.4	--	20.5
4. Coal and shale limits									
Consumption	71.6	112.1	186.9	104.4	160.0	93.8	133.0	--	--
Production	58.8	93.4	162.2	81.6	147.8	79.0	116.1	--	--
Imports	12.8	18.7	24.7	22.8	12.2	14.8	16.9	--	--
6. Nuclear moratorium, coal and shale limits									
Consumption	71.6	114.1	176.7	103.1	142.3	80.5	116.8	--	--
Production	58.8	95.2	147.8	80.2	130.1	64.1	98.0	--	--
Imports	12.8	18.9	28.9	22.9	12.2	16.4	18.8	--	--

several quantity projections reported by the longer-run models, especially in nuclear production, oil and gas production and imports, and in electricity generation. Possible reasons for these differences are explored in Sections III.4, III.5, and III.11.

The important point here to be learned from the table is that, for the most part, the FEA and DRI projections for 1990 are within the wide ranges exhibited by the longer-run models. It would therefore appear that the more basic differences in structure and inputs between the longer-run models themselves are of greater significance than the differences that may exist between the longer-run models as a group and the more detailed and disaggregated models with shorter time horizons.

III.4 LONG-RUN TRENDS IN TOTAL CONSUMPTION, DOMESTIC PRODUCTION, AND IMPORTS OF ENERGY, 1975-2010

The most consistent finding of all the models is that, barring any major restrictions on the use of nuclear or fossil-fuel technologies, domestic energy consumption and production are projected to continue to grow in the future. (Unless otherwise labeled, energy projections reported in the tables and charts refer only to domestic quantities.) Table III.3 and Figure III.1 indicate that by 2010, under the assumptions of the Base Case, and leaving aside the DESOM projections based on exogenous demand, the United States is projected to be consuming in the aggregate anywhere from 80 percent to 160 percent more energy than it consumed in 1975, and to be producing in the aggregate anywhere from 80 percent to 180 percent more energy than it produced in 1975. Furthermore, except for the ETA projections, the table shows a substantial absolute increase over 1975 levels, under the Base-Case assumptions, in the aggregate quantity of energy imports in 2010.

Translated into rates of growth, these absolute quantities imply average annual percentage increases from 1975 through 2010 of between 1.70 percent and 2.92 percent in energy consumption and between 1.70 percent and 3.12 percent in energy production. Historically, aggregate

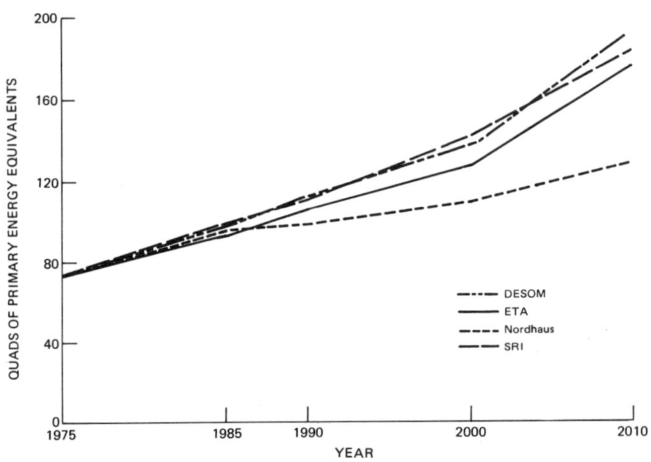

FIGURE III.1 Total energy consumption, 1975-2010, projected by the long-run models; Base Case.

energy consumption increased at a 2.5 percent annual rate over the period 1945-1975 and at a slightly lower rate of 2.4 percent over the subperiod 1965-1975.* The aggregate projections imply therefore that on *an aggregate level only*, and in the absence of policies to curtail supply or demand, U.S. consumption and production should continue to grow in the future much as it has in the past.

Despite the general agreement among the models that energy production and consumption should continue to grow, Figure III.1 and Table III.3, and subsequent charts and tables, highlight the wide range of model projections of future energy production and consumption. Throughout the remainder of Chapter III, we recognize and attempt to explain these differences, insofar as they can be explained, and where anomalies in the projections appear (as they do in some of the Nordhaus projections), face up to them in as direct a fashion as possible.

One way to put these projected energy growth rates in proper perspective is to examine the historical and projected record of GNP growth. Over the period 1945-1975, GNP increased at an annual rate of 3.4 percent; the Base-Case projections formulated by the MRG (see Chapter II) imply a slightly lower GNP growth rate of 3.19 percent. These past and projected GNP growth rates both exceed the historical growth rate of energy consumption and production and all of the projected Base-Case energy growth rates. Over the period 1975-2010, therefore, the ratio of energy consumption to GNP is projected to decline, just as this ratio has declined over the postwar period taken as a whole.† The same conclusion holds if similar calculations are performed for the Base-Case projections over the period 1975-1990 reported in Table II.2.

While the projections of aggregate energy production and consumption discussed thus far have importance for some uses, it is important to be aware of the information aggregate figures can conceal. In particular, these aggregates are reported in terms of BTU's of "primary energy equivalents," a device that attempts to put the original energy content of the various fuels in the same accounting framework. This procedure is easy to apply in the case of fossil fuels, such as coal and oil, where the "BTU content" is readily identified. It becomes somewhat more arbitrary when hydroelectric, nuclear, or solar technologies are used to replace those previously based on fossil fuels. Similar measurement difficulties arise when "synthetic" fuels are made from other fuels. Although the MRG adopted common accounting conventions in each of these cases (reported in Appendix E), the conceptual problem of mixing "apples and oranges" remains.

A second reason why an aggregate BTU figure may be misleading is that, for most purposes, one is interested in the rate of depletion of *particular* resources. Thus, to the extent one is concerned about dependence on foreign countries for oil, the relevant information is domestic

*Demand/Conservation Panel report of CONAES.

†For a fuller discussion of the record over a longer historical period, see the discussion of Energy-GNP feedback in the CONAES final report's chapter on energy demand.

production and imports of oil *specifically*, not the aggregate consumption of BTU's. Since most policy questions are fuel-specific, like the oil import issue, fuel-specific data are generally more valuable than data that are more highly aggregated.

Finally, the economist tends to be more concerned with *values* than with quantities of items that are not identical. This is because unit values, as measured by prices, are indicators of relative scarcity. In the energy context, where scarcity is the dominant concern, this suggests that value as well as quantities are important items of information. For example, on a BTU basis, oil and coal historically have sold at different prices, even during periods when no oil price controls have been in place. These prices have differed because oil and coal are not perfect substitutes in all uses. In other words, dissimilar fuels have different qualities of convenience and flexibility, and per-BTU prices reflect these variations in convenience. As a result, when the price per BTU in fuel markets is not uniform, giving equal weights to the BTU contents of all such fuels produces an aggregate BTU number having no foundation in market behavior. For many purposes, therefore, it is more instructive to examine the composition of a given aggregate of energy than to look at the aggregate quantity itself.

This is, in part, why apparent anomalies in the aggregate quantity results that appear in this and later sections of the report are not overly alarming. For example, under the Base-Case GNP growth-rate assumption, Nordhaus projects total energy consumption to rise slightly when coal and shale limits are imposed (see Table III.3). A similar result is projected by both ETA and Nordhaus when a series of constraints are added under the low GNP growth-rate assumption (scenario 9, Table III.4). Other such anomalies occur here and there throughout Chapter III.

A variety of factors can produce such oddities. The imposition of a new constraint in an optimization framework will sometimes shift production and/or consumption from earlier to later periods. Fuel-specific constraints may cause fuel shifts that, due to the accounting convention, produce increases in the total quantity of BTU's even though the level of energy services may decline. Other factors that admittedly may not be well understood are also at work. This is not to discredit the model results, but only to highlight that the state of the art and the use of the BTU accounting device can at rare times combine to produce results that are puzzling. A proposal that would alleviate these difficulties has been made by Nobay and Turvey:* to use price-weighted quantity indices of energy production or consumption similar to those used in national accounting.

Despite these defects of aggregate energy flow measures, such measures are often useful in providing rough characterizations of the effects of important policy and realization variables. Thus, Table III.3 presents the aggregate energy consumption, production, and import projections under four different policy-driven scenarios. Table III.4 supplements these projections with those for different GNP growth-rate realizations

*Ralph Turvey and A. R. Nobay, "On Measuring Energy Consumption," *Economic Journal*, December 1965, pp. 787-793.

TABLE III.4 Total Consumption, Domestic Production, and Imports of Energy in 1975, 1990, and 2010 Under Different GNP Growth-Rate Realizations (1, 7, 8) and One Strong Supply Constraint (9) (Unit: Quads of Primary Energy Equivalents)

Realizations and Policies: Base Case, Modified as Stated (Scenario Number Indicated at Left)		DESOM		ETA		Nordhaus		SRI	
	1975	1990	2010	1990	2010	1990	2010	1990	2010
1. Base Case									
Consumption	71.6	112.5	199.1	104.8	178.9	96.2	129.6	108.9	185.4
Production	58.8	93.6	171.2	81.5	167.0	81.1	106.7	94.7	166.6
Imports	12.8	18.9	25.1	23.3	11.9	15.1	22.9	14.2	18.8
7. High GNP growth rate									
Consumption	71.6	--	--	123.8	256.4	113.3	199.9	--	293.6
Production	58.8	--	--	97.6	246.2	96.0	175.1	--	272.0
Imports	12.8	--	--	26.2	10.2	17.3	24.8	--	21.6
8. Low GNP growth rate									
Consumption	71.6	--	--	92.7	98.8	81.6	82.3	--	117.8
Production	58.8	--	--	71.6	83.7	68.1	67.9	--	99.7
Imports	12.8	--	--	21.1	15.1	13.5	14.4	--	18.1
9. Low GNP growth rate, nuclear moratorium, and coal and shale limits									
Consumption	71.6	--	--	91.0	99.5	72.0	83.0	--	--
Production	58.8	--	--	70.2	86.3	56.3	70.0	--	--
Imports	12.8	--	--	20.8	13.2	15.7	13.0	--	--

and, in the case of the last scenario in that table, for low GNP growth in the presence of two supply-restrictive policies imposed simultaneously.

Taking the results in Table III.3 first, the principal finding from all of the scenarios is that neither a policy that rules out further nuclear plant construction (scenario 3) nor one that imposes the "medium" coal and shale limits defined in Section II.2a (scenario 4) appear by themselves to cut down aggregate consumption or production materially.* These are logical consequences of the input assumptions and model structures that lead to a shift from any restricted technology to its closest substitute(s). However, the impositon of both types of constraints results in a cut in energy production substantially more severe than the sum of the cuts induced by either restriction taken separately. Sections III.5 and III.7 discuss in greater detail the substitutability of coal and nuclear power, and how energy prices are projected to respond to restrictions placed on coal and/or nuclear power.

An interesting aspect of the policy scenario results reported in Table III.3 is the behavior of imports. All of the models project a significant role for imports throughout the 1975-2010 period. Moreover, except for the Nordhaus projections, the magnitude of imports in both 1990 and 2010 does not vary materially between scenarios. In part, these outcomes can be attributed to the MRG-imposed import constraint, which limits oil and gas imports annually to 33 percent of domestic consumption, and cumulatively, to 1,000 Quads. The constraint essentially "spreads out" a limited quantity of oil imports over the projection period in a manner that is not highly sensitive to possible restrictions on other technologies or energy resources. As explained in Chapter II, this constraint reflects both policy and realization considerations.

Another interesting set of results is contained in Table III.4, which presents the aggregate energy consumption, production, and import projections under four alternative realizations. Aside from the Base Case, these include the MRG high and low GNP growth paths and the low GNP growth path in the presence of nuclear, coal, and shale restrictions.

To assess how these projections are related to GNP growth, Table III.5 illustrates the growth rates over the 1975-2010 period in GNP and energy consumption and production implied by the MRG assumptions and projections. Together, Tables III.4 and III.5 demonstrate that GNP growth is one of the most important driving variables in the study. In particular, the projections in these tables indicate that a variation in the annual rate of GNP growth of approximately 2.8 percentage points can, under MRG assumptions, and in the absence of policies constraining supply, produce differences in aggregate energy consumption in 2010 of between 117.6 Quads (Nordhaus) and 157 Quads (ETA), amounts far in excess of the *total* energy consumption in 1975 of 71.6 Quads.

*DESOM projections in scenarios of "low" coal and shale oil limits (see Section II.2a) reinforce the conclusions stated in this paragraph. As stated in Section II.2a, unless otherwise defined, the policy of "coal and shale oil limits" refers specifically to the "moderate" limits on coal and shale oil set forth in that section.

TABLE III.5 Rates of Growth, in Percent per Annum, 1975-2010, in GNP and Total Energy Consumption and Production, Under Different GNP Growth-Rate Realizations and One Policy Constraint

Realizations and Policies: Base Case, Modified as Stated (Scenario Number Indicated at Left)	DESOM	ETA	Nordhaus	SRI
1. Base Case				
GNP	3.19	3.19	3.19	3.19
Consumption	2.92	2.65	1.71	2.76
Production	3.12	3.03	1.72	3.02
7. High GNP growth				
GNP	4.65	4.65	4.65	4.65
Consumption	--	3.71	2.98	--
Production	--	4.18	3.17	--
8. Low GNP growth				
GNP	1.78	1.78	1.78	1.78
Consumption	--	0.94	0.40	--
Production	--	1.06	0.41	--
9. Low GNP growth, nuclear moratorium, coal and shale limits				
GNP	1.78	1.78	1.78	1.78
Consumption	--	0.94	0.42	--
Production	--	1.10	0.50	--

The projections by all of the models also demonstrate that, in all of the realization scenarios, both energy consumption and production grow at slower rates than GNP. Thus, in the Base Case, although GNP grows at an annual rate of 3.19 percent, energy production and consumption are projected to grow at lower rates by all of the models, and at substantially lower rates by Nordhaus. A similar pattern exists in each of the other three scenarios.

Finally, through its earlier work, the MRG performed an extensive sensitivity analysis of all of these aggregate energy projections with respect to differences in realizations. Among these sensitivity scenarios, the MRG investigated the effects of low uranium availability, of the absence of non-price-induced conservation, of relaxation of current environmental restrictions on coal and nuclear technologies, and of different cost assumptions for various technologies. Without exception, the differences in aggregate energy production due to these changes in realizations fell well within the range of results presented here with respect to differences in GNP growth rates.

The one variable that is capable of producing especially large differences in the aggregate estimates, however, is the price elasticity of demand. As explained in Chapter II, the MRG was unable to standardize price and income elasticities across models. Section III.11 discusses how many of the differences in model results can be traced to differences in elasticities. That discussion includes results of parametric variations of the aggregate price elasticity built into the ETA model, in particular. An advance indication is provided here by Table III.3, which reports the output results from all of the long-run models. As explained in Appendixes A and C, end-use demands for energy services are not responsive to price changes in the DESOM model, in contrast with the nonzero price elasticities incorporated in the other models. As a result, the DESOM projections of aggregate energy production and consumption in 2010 are consistently higher and tend to exhibit relatively less variation over the broad range of scenarios than is the case with the other models. In those models, price elasticity differences are important to the projections of aggregate energy consumption. Subsequent comparison of the DESOM results with the results of the other models will show that the effects of elasticity differences are just as important on a more disaggregated level.*

Subsequent parts of Chapter III present and explain the MRG findings in a much more disaggregated fashion.

III.5 COMPOSITION OF DOMESTIC ENERGY PRODUCTION, 1975-2010

Throughout the twentieth century, the United States has relied almost exclusively on fossil fuels for its energy needs. The principal aim of the MRG is to use the available energy models to project what future mixes of fuels might satisfy the energy needs of the United States during the next 35 years under a variety of policy and realization scenarios.

In particular, each of the models projects under the MRG assumptions what energy supply configuration would meet energy demands at minimum cost over time. There are a series of factors, however, that may act to place limits on the use of certain technologies. For example, the difficulties of placing values on environmental damage, federal and state licensing requirements, possible labor and capital shortages in certain energy sectors, and potential water shortages all may, to varying degrees, constrain the use of coal, shale oil, and nuclear power. As a result, the MRG specified scenarios in which partial or total restrictions are imposed on each of these technologies, both separately and in combination (see Chapter II). In addition, variations in the growth rate of GNP were used to project the composition of energy supply under different demand conditions.

Together, these projections indicate that the next 35 years should be quite different from the last 35. Oil and natural gas, our most important energy sources during most of this century, are projected to decline

*Elasticity differences are also important in explaining the differences between the ETA and Nordhaus projections. See Section III.11.

in importance relative to other fuels. Coal production, on the other hand, is projected in all of the scenarios considered by the MRG to increase substantially, perhaps even at growth rates that may strain the supply of labor potentially available to the industry. Most significantly, however, the 2010 projections envision substantial roles for both nuclear power (except in scenarios of nuclear moratorium) and synthetic fuels, technologies at present of relatively minor and no importance, respectively. Finally, other energy sources such as geothermal and solar energy are projected to provide some of our energy in 2010.

The strength of each of these conclusions depends, of course, on future policies and realizations. More generally, the same is true for the overall composition of domestic energy production between now and 2010 and beyond. In this section, therefore, we present the MRG projections of the composition of energy supply under the nine scenarios specified in Chapter II. These results will be supplemented by additional information gained through sensitivity analyses of these basic scenarios.

The discussion begins with Figure III.2 and Table III.6A, both of which present the composition projections for the Base Case. All of the models project (in Figure III.2) rising coal and nuclear production and gradually declining oil and natural gas production over the 1975-2010 period. Whereas in 1975, when approximately 25 percent, 3 percent, and 68 percent of our energy were produced by coal, nuclear, and oil and natural gas, respectively, the models project that in 2010, under Base-Case assumptions, these percentages may look more like 51 percent, 31 percent, and 16 percent.* Thus, from an economy that is now heavily dependent on oil and natural gas, the Base-Case projections envision an economy in 2010 that relies primarily on coal and nuclear power.

By 2010, we may also see the use of other technologies or energy sources that are now of relatively minor or no importance. DESOM, ETA, and SRI all project a growth of hydroelectric and geothermal energy over the period and the introduction of shale oil technology after 2000. The growth of the latter is limited by the MRG constraint on the rate of capacity expansion for all new technologies (see Chapter II). DESOM and SRI also project the introduction of solar heating and cooling in 1985, and the expansion of demand for these purposes to between 3.0 and 3.2 Quads of production by 2010.†

There is nothing sacrosanct about the Base-Case result, however. A variety of policies may be imposed between now and 2010 that could, by fiat, alter the supply composition throughout the period for environmental or other reasons. Table III.6A contains supply composition projections for 2010 under three such policies. These are supplemented by Figures III.3 (for DESOM), III.4 (for ETA), and III.5 (for Nordhaus),

*These figures represent arithmetic averages of the Base-Case projections of the four models. See Table III.6A for the underlying data.

†These models do not project the introduction of solar-produced electricity, however, at any time during the 1975-2010 period, under the Base-Case assumptions.

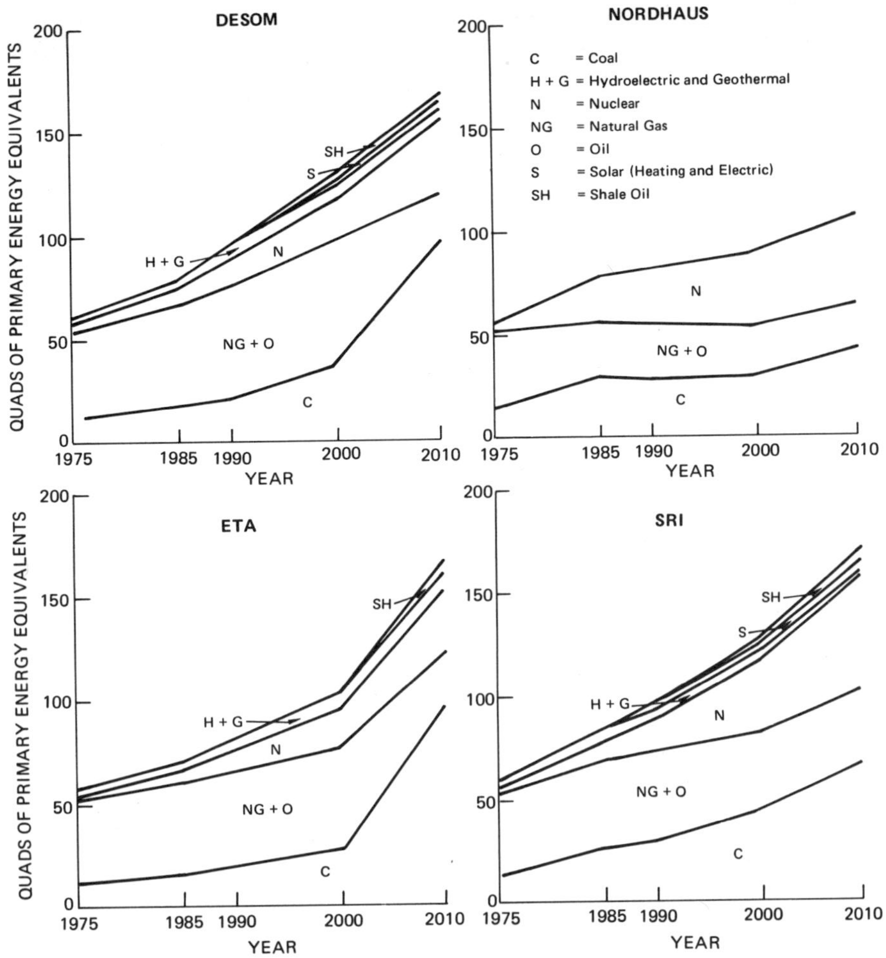

FIGURE III.2 Composition of total energy production, 1975-2010, Base Case. NOTE: In DESOM and SRI, direct solar is separate from AES (see text).

which illustrate the composition projections over time for four different scenarios. Three of these scenarios are represented in Table III.6A and one in Table III.8. In reading these charts and tables, one should note that in DESOM and SRI direct solar heating and cooling is combined with solar electric, and hence is not included in the definition of the AES for those models. In Nordhaus and ETA, however, direct solar *is* included in the AES, while solar electric is in a category of its own.

In comparing the results of these scenarios, one should keep in mind that in the Base Case the models project heavy reliance on both coal and nuclear energy by 2010. Policies that impinge on only one of these energy sources, such as a nuclear moratorium or limits on coal and shale production, should therefore affect only part of the future energy network. Such policies are to be contrasted with those, such as the *simultaneous* imposition of nuclear, coal, and shale restrictions, which should affect the bulk of energy production by 2010.

TABLE III.6A Composition of Total Domestic Energy Production, 2010, Under Different Policy Scenarios (Unit: Quads of Primary Energy Equivalents)

Policies: Base Case, Modified as Stated (Scenario Number Indicated at Left)	DESOM	ETA	Nordhaus	SRI
1. Base Case				
Total production	*171.2*	*167.0*	*106.7*	*166.6*
Coal	99.8	98.6	41.6	78.2
Natural gas, oil, and NGL[a]	22.7	23.7	23.3	27.1
Nuclear	35.4	31.2	41.8	52.8
Solar (heating and electric)	3.0	0.0	0.0	3.2
Shale oil	4.1	6.0	0.0	1.8
Hydroelectric and geothermal	6.2	7.5	0.0	3.8
Nonelectric AES	0.0	0.0	0.0	0.0
Other	0.0	0.0	0.0	0.2
3. Nuclear moratorium				
Total production	*166.9*	*163.6*	*103.2*	*147.8*
Coal	127.7	126.4	81.3	106.1
Natural gas, oil, and NGL	31.8	23.6	21.8	30.4
Nuclear	0.0	0.1	0.0	0.1
Solar (heating and electric)	1.0	0.0	0.0	3.4
Shale oil	0.2	6.0	0.0	3.1
Hydroelectric and geothermal	6.2	7.5	0.0	4.1
Nonelectric AES	0.0	0.0	0.0	0.0
Other	0.0	0.0	0.0	0.6
4. Coal and shale limits				
Total production	*162.2*	*147.8*	*116.1*	--
Coal	39.3	45.0	62.8	--
Natural gas, oil, and NGL	22.7	24.5	23.0	--
Nuclear	57.4	41.0	30.3	--
Solar (heating and electric)	3.0	0.0	0.0	--
Shale oil	5.9	6.0	0.0	--
Hydroelectric and geothermal	6.2	7.5	0.0	--
Nonelectric AES	27.7	23.8	0.0	--
Other	0.0	0.0	0.0	--
6. Nuclear moratorium and coal and shale limits				
Total production	*147.8*	*130.1*	*98.0*	--
Coal	44.3	45.0	62.8	--
Natural gas, oil, and NGL	18.2	24.4	25.0	--
Nuclear	0.0	2.3	0.0	--
Solar (heating and electric)	8.4	1.4	0.0	--
Shale oil	5.9	6.0	0.0	--
Hydroelectric and geothermal	7.2	7.5	0.0	--
Nonelectric AES	63.8	43.6	10.1	--
Other	0.0	0.0	0.0	--

[a] Natural gas liquids.

NOTE: In DESOM and SRI, direct solar is separate from the AES.

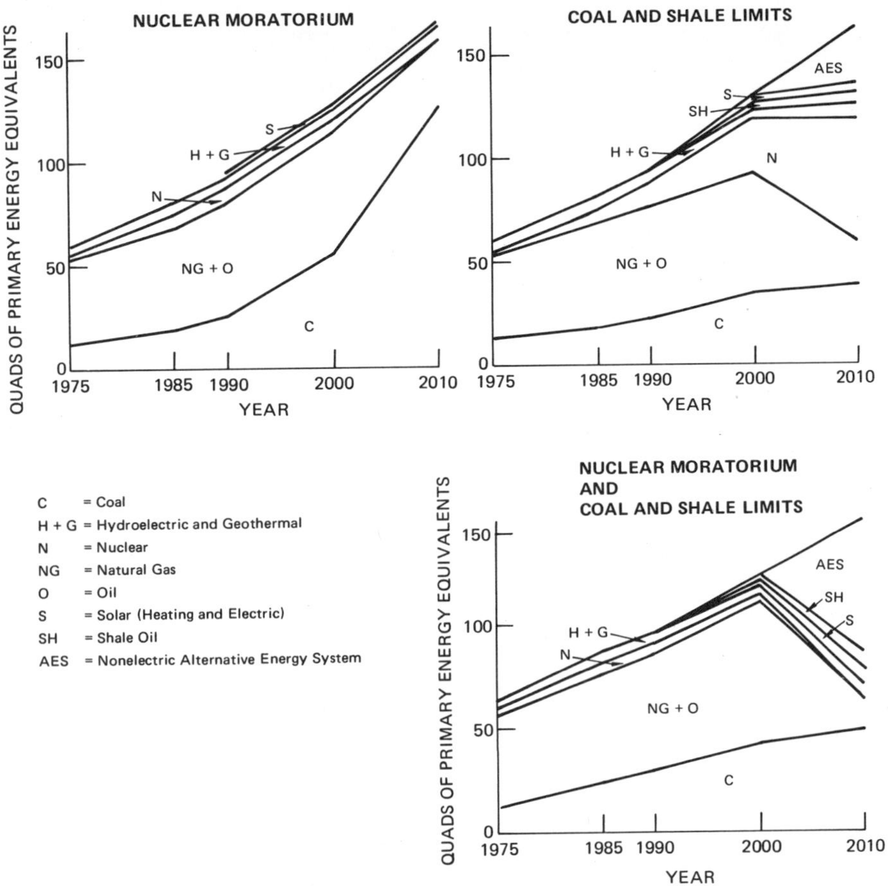

FIGURE III.3 Projections by DESOM of the composition of total energy production, 1975-2010, under four different scenarios. NOTE: In DESOM, direct solar is separate from the AES.

One way to measure the differences in the effects of these policies is to examine the effects on prices, as we do in Section III.7. Another way is to examine the differences in real income that may be generated, as we do in Section III.8. In this section, however, we look at the differences in supply composition in an effort to discover what energy sources the energy supply system is projected to turn to if one, or perhaps two, energy sources and technologies are restricted or rendered unavailable by deliberate policy choice.

The tables and charts illustrate that, to differing degrees, coal and nuclear power are substitutes. In the nuclear moratorium scenario, all of the models project that in 2010 nearly the entire reduction in nuclear production is offset by expanded coal production (compare coal and nuclear production results for scenarios 1 and 3 in Table III.6A). Figures III.3 through III.5 illustrate these outcomes for the period between 1975 and 2010.

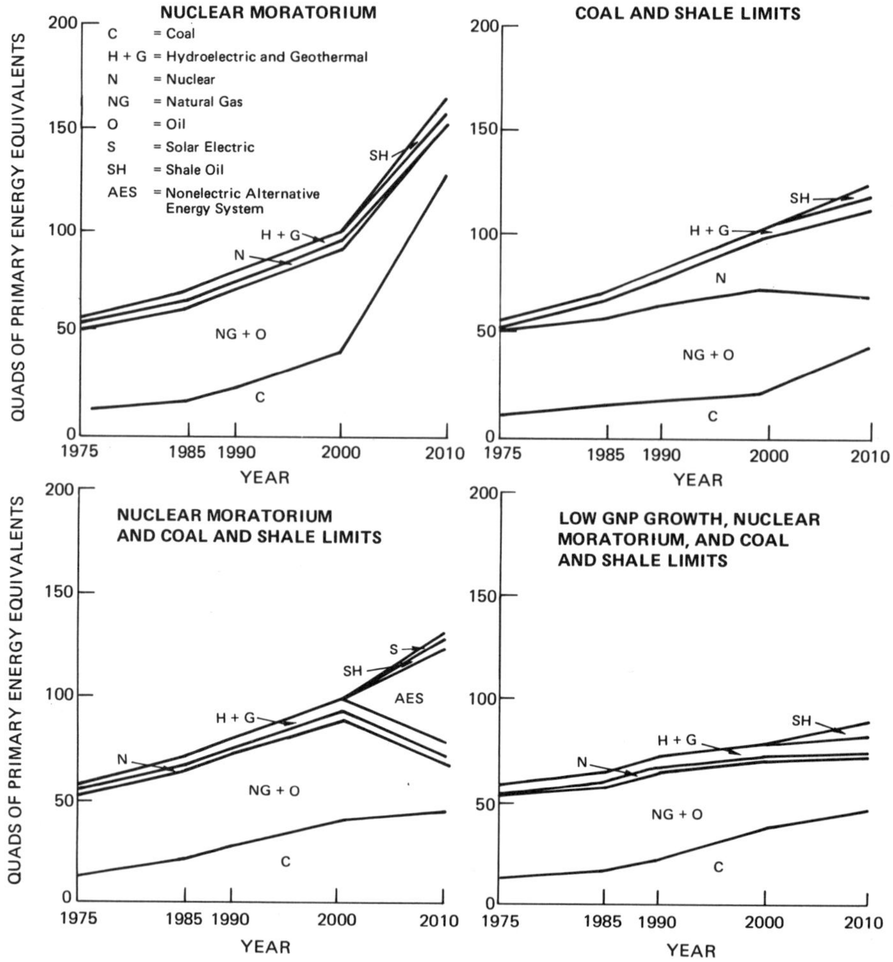

FIGURE III.4 Projections by ETA of the composition of total energy production, 1975-2010, under four different scenarios.

Projections for the coal and shale limits case yield corresponding results, but with some difference in degree. ETA projects that in 2010 the 53.6-Quad reduction (104.6-49.0) in coal and shale production resulting from the imposition of the limits will be distributed among increased nuclear production (9.8 Quads), a substantial rise in the use of an AES (23.6 Quads), a slight rise in oil and gas production (0.9 Quads), and a drop in overall production (19.2 Quads). A similar projection is made by DESOM: the 60.5-Quad decline in coal production in 2010 is distributed among a 22.0-Quad increase in nuclear production, a rise of 27.7 Quads in AES production, a slight rise of 1.8 Quads of shale oil output, and an overall drop in production of 8.9 Quads. The Nordhaus projections for 2010 for the coal and shale limits case are confusing in that coal production is projected to be 21.2 Quads *higher* than in the Base Case. This anomaly is attributable to a different method of implementing the

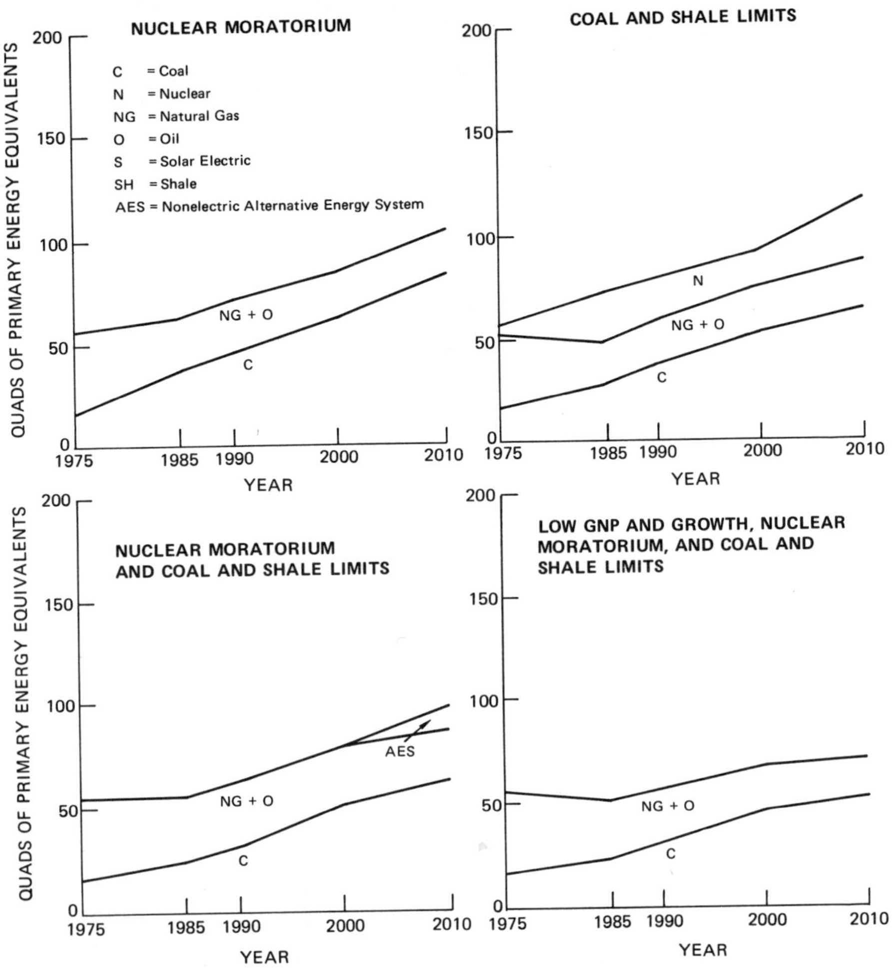

FIGURE III.5 Projections by Nordhaus of the composition of total energy production, 1975-2010, under four different scenarios.

coal and shale limits constraint in Nordhaus than in ETA.* Hence the Nordhaus simulations, although they complicate the results, are not inconsistent with the basic findings by DESOM and ETA: that in the presence of coal and shale oil restrictions, nuclear power is, to a limited extent, a substitute for coal.

*Nordhaus imposed an asymptotic limit of 62.8 Quads on coal and shale oil *combined*; ETA imposed an asymptotic limit of 50 Quads on coal and 12 Quads on shale oil *separately*. Moreover, the *rate of growth* of coal and shale oil production was constrained in Nordhaus, whereas in ETA the actual quantity of coal and shale oil production in each time period was constrained. This led to an implicit subsidy in the Nordhaus model on coal production in 2010. The effect on other years and on the value of the objective function, however, is minor.

The first conclusion, therefore, to emerge from the policy scenarios is that coal is a much better substitute for nuclear power than nuclear power is for coal. In part, this conclusion rests on the assumption that effective technologies for gasifying and liquefying coal will be developed at the Base-Case costs used by the MRG. Substantially higher synthetic fuels costs, however, could significantly reduce the substitutability of coal for nuclear power. Otherwise, the substitutability result is not a surprising one, for the only use that coal and nuclear energy have in common is in the generation of electricity. Thus, when a "moratorium" is imposed on nuclear power, whose only use is in the generation of electricity, coal stands as a ready substitute. But when limits are placed on coal and shale, nuclear power cannot substitute for them as a base fuel for the production of synthetic oil or gas. Nevertheless, the imposition of *either* nuclear or coal and shale restrictions does not seem to place a heavy strain on the ability of the energy system to adapt, as measured by the projected increase in the production of alternate energy sources.

An entirely different conclusion is projected by the models when both a nuclear moratorium and limits on coal and shale oil are imposed. From a 2010 Base-Case supply configuration in which coal and nuclear power are projected to comprise approximately three-fourths of total energy production, the imposition of both these supply constraints causes a reduction in total production (of between 8 percent and 22 percent), as well as a movement toward the adoption of other energy sources. DESOM and ETA, for example, project the adoption of a clean, nonelectric Alternative Energy System (AES) in 2000 (see Figures III.3 and III.4), which DESOM projects to supply up to 40 percent of total energy consumption in 2010. As indicated in Chapter II, the AES may be one or more of many energy supply technologies which in the Base Case can provide energy at a cost of $\$5/10^6$ BTU. Nordhaus, too, projects the adoption of an AES by 2000, but indicates a smaller role for it in 2010 than projected by either DESOM or ETA.

Despite these numerical differences, Figures III.3 through III.5 clearly indicate a need for an AES from the end of this century on if policies involving both nuclear moratorium and limits on coal and shale oil production are to be enacted. In that case, nuclear power would be unavailable for electricity generation, while the capital cost of solar-produced electricity would (with Base-Case assumptions) be over three times as high as coal-fired plants. As a result, the limited flow of coal allowed by the policy would be channeled into electricity generation, leaving two alternatives for nonelectric demands: (1) that they be filled by a clean nonelectric AES (e.g., direct solar,* cogeneration, new oil discoveries, or as yet undiscovered energy supply technologies), or (2) that they be sharply reduced by substantial price increases. To some extent, (2) occurs with (1) since the Base-Case cost of the AES is assumed to be $\$5/10^6$ BTU, roughly 2.5 times the current domestic price of crude oil. Without the AES, however, prices would certainly rise to

*Considered separately from the AES in DESOM.

substantially higher levels. In the end, the models indicate that if we as a society intend to restrict nuclear, coal, and shale oil technologies, and if the actual outcomes are analogous to, or have the same effect as, those simulated by the MRG, then it would be highly prudent now to explore seriously whether and how a clean nonelectric AES can be developed and operated at relatively low cost.

The need for a low-cost AES becomes even stronger if the "low coal and shale limits" defined in Chapter II are implemented. Table III.6B contrasts the DESOM supply composition projections for the two coal and shale limits scenarios, with and without a nuclear moratorium. In each case, the imposition of substantially lower limits on the production of coal and shale oil produces much greater reliance on the AES. In particular, when both the low coal and shale limits and the nuclear moratorium are imposed, the AES supplies approximately 70 percent (103 ÷ 142.5) of total domestic energy production.

The DESOM projections, however, are a type of limiting case, because demands for energy services are assumed to be insensitive to the substantial price increases (see Section III.7) that would be generated by

TABLE III.6B DESOM Projections of Domestic Energy Production, 2010, Under Different Policy Scenarios (Unit: Quads of Primary Energy Equivalents)

	Mod. C&S Limits[a]	Mod. C&S Limits[a] and Nuclear Moratorium	Low C&S Limits[b]	Low C&S Limits[b] and Nuclear Moratorium
Total Production	*162.2*	*147.8*	*153.8*	*142.5*
Coal	39.3	44.3	21.4	21.4
Natural gas, oil, and natural gas liquids	22.7	18.2	22.7	0.0
Nuclear	57.4	0.0	54.4	0.0
Solar (heating and electric)	2.0	8.4	1.4	8.3
Shale oil	5.9	5.9	3.6	3.6
Hydroelectric and geothermal	6.2	7.2	6.2	6.2
Nonelectric AES	27.7	63.8	44.1	103.0
Other	0.0	0.0	0.0	0.0

[a]Mod. C&S Limits (Moderate Coal and Shale Limits): 2 and 40 Quads of coal and shale oil production, respectively, in 2000, and ultimately 12 and 50 Quads of coal and shale oil, respectively.

[b]Low C&S Limits (Low Coal and Shale Limits): 25 Quads of coal and shale oil combined.

NOTE: In DESOM, direct solar is separate from the AES.

a low coal and shale limits policy. Nevertheless, they provide dramatic evidence of the need for low-cost, clean energy sources in the event society depletes its oil and natural gas and chooses not to use extensively its nuclear and coal energy sources for environmental or other policy reasons.

The foregoing conclusions are not significantly altered if we explicitly recognize the possibility of a "moratorium" applied only to the Fast Breeder Reactor (FBR) and all Advanced Converter Reactors (ACR's). Table III.7 illustrates the ETA and Nordhaus projections of the supply

TABLE III.7 Composition of Total Energy Production, 2010 and 2030, Comparing Two Policy Scenarios, With and Without Advanced Nuclear Reactors (Units: Quads of Primary Energy Equivalents)

Policies: Base Case, Modified as Stated (Scenario Number Indicated at Left)	ETA		Nordhaus	
	2010	2030	2010	2030
1. Base Case				
Total	*167.0*	*258.4*	*106.7*	*188.7*
Coal	98.6	192.0	41.6	104.7
Oil, natural gas, and NGL	23.7	3.5	23.3	15.5
Nuclear	31.2	39.1	41.8	68.7
All other	13.5	23.8	0.0	0.0
2. Moratorium on all ACR's and the FBR				
Total	*166.5*	*257.8*	*106.7*	*188.1*
Coal	98.4	192.2	41.6	129.5
Oil, natural gas, and NGL	23.7	3.5	23.3	15.5
Nuclear	30.9	38.2	41.8	37.2
All other	13.5	23.8	0.0	1.9
4. Coal and shale limits				
Total	*147.8*	*219.4*	*116.1*	*165.9*
Coal	45.0	49.0	62.8	62.8
Oil, natural gas, and NGL	24.5	3.8	23.3	15.8
Nuclear	41.0	80.7	30.0	72.7
Other	37.3	85.9	0.0	0.0
5. Coal and shale limits, moratorium on all ACR's and the FBR				
Total	*146.6*	*213.1*	*114.2*	*161.5*
Coal	45.0	49.0	62.8	62.8
Oil, natural gas, and NGL	24.5	3.8	23.3	15.8
Nuclear	39.8	71.5	28.1	66.5
All other	37.3	88.8	0.0	16.5

composition in 2010 and 2030 under four different scenarios, two with and two without the FBR and all ACR's. For the year 2010, the table indicates virtually identical results in each model whether or not the advanced nuclear technology is permitted. (Compare results for scenario 1 with 2, and those for 4 with 5.)

By the year 2030, however, both model projections suggest that the composition of energy supply could be moderately affected by the presence or absence of a moratorium on all ACR's and the FBR. The magnitude of this effect will depend, of course, on whether it would be cheaper, in the absence of a moratorium, to use such advanced nuclear reactors. Nordhaus projects some use of an ACR in the Base Case by 2010 (see Table III.12A below for a breakdown of the nuclear totals presented in Table III.7) and hence its elimination prompts a much greater reliance on coal by 2030. When coal and shale limits are already in place, Nordhaus projects the substitution of another technology for the barred advanced reactors, specifically a nonelectric AES. ETA, on the other hand, shows only moderate supply composition effects from a moratorium on the advanced reactors since it projects neither type of reactor to be in use unless coal and shale limits are in place. The role of the advanced reactors in our energy supply network will be discussed in more detail toward the end of this Section III.5.

Up to this point only policy scenarios have been described. Table III.8 presents projections of supply composition under three different GNP realizations and one additional policy. Most significantly, all models project a positive correlation in coal and nuclear production in response to higher or lower GNP growth. They disagree, however, on the relationship between oil and gas production in 2010 and GNP growth; Nordhaus and SRI project a positive relationship and ETA projects a negative relationship between the two variables for that year. This difference is primarily due to differences in the time pattern of oil and gas production, however, rather than in the total amounts produced over the period 1975-2010.

An important component in all of the projections discussed thus far is coal production, even in the scenarios in which coal and shale oil limits are imposed. One of the reasons for this is that even where nuclear power is permitted, coal is also used to generate electricity. This is more true in ETA (which restricts the introduction of nuclear plants to no more than 60 percent of the expansion in total electricity-generating capacity) than in Nordhaus, but it occurs in both. Similarly, coal will continue to be burned directly in industrial uses, although the models also differ in the way direct coal use is projected into the future.*

Potentially the largest use for coal in the future, however, is in the manufacture of synthetic fuels. Tables III.9A and III.9B present projections by DESOM and ETA of coal used, by function, in a variety of policy and realization scenarios. The projections are similar to a remarkable degree. In all but the coal and shale oil limits scenarios

*See Section III.11 and Appendix D for further discussion of this issue.

TABLE III.8 Composition of Total Energy Production, 2010, Under Different Realization Scenarios and One Policy Constraint (Unit: Quads of Primary Energy Equivalents)

Realizations and Policies: Base Case Modified as Stated (Scenario Number Indicated at Left)	DESOM	ETA	Nordhaus	SRI
1. Base Case				
Total production	*166.7*	*167.0*	*106.7*	*166.6*
Coal	109.4	98.6	41.6	78.2
Natural gas, oil, and NGL	18.7	23.7	23.3	27.1
Nuclear	28.0	31.2	41.8	52.8
Solar (heating and electric)	0.0	0.0	0.0	3.2
Shale oil	3.7	6.0	0.0	1.8
Hydroelectric and geothermal	6.1	7.5	0.0	3.8
Nonelectric AES	0.8	0.0	0.0	0.0
Other	0.0	0.0	0.0	0.2
7. High GNP growth				
Total production	--	*246.2*	*175.1*	*266.2*
Coal	--	121.3	83.3	136.6
Natural gas, oil, and NGL	--	20.3	28.2	30.3
Nuclear	--	55.0	55.1	87.1
Solar (heating and electric)	--	0.0	0.0	4.4
Shale oil	--	6.0	3.7	3.3
Hydroelectric and geothermal	--	7.5	0.0	3.8
Nonelectric AES	--	36.1	4.8	0.0
Other	--	0.0	0.0	0.7
8. Low GNP growth				
Total production	--	*83.7*	*67.9*	*99.7*
Coal	--	24.7	22.9	36.6
Natural gas, oil, and NGL	--	30.2	18.9	25.5
Nuclear	--	15.3	26.0	30.3
Solar (heating and electric)	--	0.0	0.0	2.5
Shale oil	--	6.0	0.0	1.0
Hydroelectric and geothermal	--	7.5	0.0	3.8
Nonelectric AES	--	0.0	0.0	0.0
Other	--	0.0	0.0	0.1
9. Low GNP growth, nuclear moratorium, and coal and shale limits				
Total production	--	*86.3*	*70.0*	--
Coal	--	45.0	50.8	--
Natural gas, oil, and NGL	--	26.5	19.1	--
Nuclear	--	1.3	0.0	--
Solar (heating and electric)	--	0.0	0.0	--
Shale oil	--	6.0	0.0	--
Hydroelectric and geothermal	--	7.5	0.0	--
Nonelectric AES	--	0.0	0.0	--
Other	--	0.0	0.0	--

NOTE: In DESOM and SRI, direct solar is separate from the AES.

TABLE III.9A DESOM Projections of Coal Used, by Function, 1975, 1990, 2010, and 2020, Under Different Policy Scenarios (Unit: Quads of Primary Energy Equivalents)

Policies: Base Case, Modified as Stated (Scenario Number Indicated at Left)	1975	1990	2010	2020
1. Base Case				
Total coal production	*13.3*	*20.5*	*99.8*	*165.1*
Coal for electricity generation	8.8	12.7	19.2	31.8
Coal for synthetic fuels	0	0.4	69.3	121.0
Coal for direct and other uses	4.5	7.4	11.3	12.3
3. Nuclear moratorium				
Total coal production	*13.3*	*25.4*	*127.7*	*212.3*
Coal for electricity generation	8.8	17.6	51.1	72.2
Coal for synthetic fuels	0	0.4	65.4	127.8
Coal for direct and other uses	4.5	7.4	11.2	12.3
4. Coal and shale limits				
Total coal production	*13.3*	*22.1*	*39.3*	*44.2*
Coal for electricity generation	8.8	14.2	1.5	1.4
Coal for synthetic fuels	0	0.5	26.6	30.5
Coal for direct and other uses	4.5	7.4	11.2	12.3
6. Nuclear moratorium, coal and shale limits				
Total coal production	*13.3*	*24.9*	*44.3*	*49.8*
Coal for electricity generation	8.8	17.5	37.7	37.5
Coal for synthetic fuels	0	0	0	0
Coal for direct and other uses	4.5	7.4	6.6	12.3

and the low-GNP-growth scenario, synthetic fuels are the dominant claimant on coal production in 2010. In the next year recorded (2020 for DESOM, 2030 for ETA), they require the largest use of coal in all scenarios except where coal and shale oil limits are imposed simultaneously with a nuclear moratorium.

The large absolute quantities of coal production projected by DESOM and ETA and, particularly, the projections of coal used for synthetic fuels production, raise some important issues. First, some may question whether the coal mining industry can attract sufficient labor and capital to grow at a rate which, in the Base-Case ETA projections, would increase total coal production sevenfold. In response, much if not all of the coal projected for synthetic fuels production represents western coal, which is strip-mined, and therefore, not labor-intensive. As to a

TABLE III.9B ETA Projections of Coal Used, by Function, 1975, 1990, 2010, and 2030, Under Different Policy and Realization Scenarios (Unit: Quads of Primary Energy Equivalents)

Policies and Realizations: Base Case, Modified as Stated (Scenario Number Indicated at Left)	1975	1990	2010	2030
1. Base Case				
Total coal production	*13.3*	*20.1*	*98.6*	*192.0*
Coal for electricity generation	8.8	14.0	29.6	37.1
Coal for synthetic fuels	0	0	60.0	141.5
Coal for direct and other uses	4.5	6.1	9.0	13.4
3. Nuclear moratorium				
Total coal production	*13.3*	*25.0*	*126.4*	*228.1*
Coal for electricity generation	8.8	18.9	57.4	71.1
Coal for synthetic fuels	0	0	60.0	143.6
Coal for direct and other uses	4.5	6.1	9.0	13.4
4. Coal and shale limits				
Total coal production	*13.3*	*19.4*	*45.0*	*49.0*
Coal for electricity generation	8.8	13.3	19.1	8.1
Coal for synthetic fuels	0	0	16.9	27.4
Coal for direct and other uses	4.5	6.1	9.0	13.4
6. Nuclear moratorium, coal and shale limits				
Total coal production	*13.3*	*28.1*	*45.0*	*49.0*
Coal for electricity generation	8.8	22.0	36.0	35.6
Coal for synthetic fuels	0	0	0	0
Coal for direct and other uses	4.5	6.1	9.0	13.4
7. High GNP growth				
Total coal production	*13.3*	*25.1*	*121.3*	*430.0*
Coal for electricity generation	8.8	19.0	52.3	85.3
Coal for synthetic fuels	0	0	60.0	331.3
Coal for direct and other uses	4.5	6.1	9.0	13.4
8. Low GNP growth				
Total coal production	*13.3*	*17.5*	*24.7*	*68.5*
Coal for electricity generation	8.8	11.4	14.5	12.4
Coal for synthetic fuels	0	0	1.1	42.8
Coal for direct and other uses	4.5	6.1	9.0	13.4

possible capital shortage, Section III.10 indicates that such an outcome is unlikely.

The second, and potentially more important, issue is whether coal production at levels projected by DESOM, ETA, and SRI for most of the scenarios will be environmentally acceptable. In particular, in the absence of any limits on coal production, both ETA and SRI project coal use by 2010 to be six to eight times present production. Without considerable improvements in pollution-abatement technology, such levels of coal use and production could pose serious environmental risks in the United States and add to a possible worldwide CO_2 problem.

It was in response to this concern that the MRG decided to consider a policy that placed limits on coal and shale oil production. But whether society will choose to adopt such constraints or prefer to live with projected high levels of coal and shale production is an open question. The answer will, of course, depend on improvements in land reclamation techniques (for strip-mined coal) and emissions removal technologies (for both coal burning and synthetic fuels manufacture), on solution of water availability problems (for shale oil), and on public attitudes in the future. Whatever the outcome, the coal and shale projections reported here indicate that the issue will have to be faced.

A partial qualification to the foregoing remarks is that Nordhaus projects considerably lower coal production than the other models in all scenarios (see Tables III.7 and III.8). As Section III.11 indicates, this is because of differences in the way the Nordhaus model treats cross-elasticities of demand between coal and other fuels. Consequently, the projections of synthetic fuels production made by other models are substantially greater than those made by Nordhaus.* Nevertheless, the Nordhaus projections reinforce those made by the other models, which indicate a substantial increase in the amount of coal production in nearly all scenarios.

In addition to projecting the primary energy composition of total energy production, the MRG also projected the primary energy composition of electricity generation under a variety of scenarios. Table III.10 presents such projections for four policy scenarios for both 1990 and 2010. Since the projections by all of the models except SRI do not account for regional differences in energy costs and demands, these projections are intended to provide only a rough indication of how the nationwide percentage composition of electricity generation might vary under different policy scenarios.

All models project increasing reliance on nuclear power in the Base Case. As remarked earlier, ETA projects a slower shift to nuclear because of a built-in constraint. This constraint is designed to account for regional differences and load-curve characteristics, which combine to limit the speed of introduction of nuclear-powered electricity-generating facilities. In contrast, Nordhaus tends to report higher

*The functional breakdown of coal production, as presented in Tables III.9A and III.9B, could not easily be accomplished for either the Nordhaus or the SRI projections; hence, they are not shown here.

TABLE III.10 Projections of Percentage Composition of Electricity Generation in 1990 and 2010, Under Different Policy Scenarios[a]

Policies: Base Case, Modified as Stated (Scenario Number Indicated at Left)		DESOM		ETA		Nordhaus		SRI	
	1975	1990	2010	1990	2010	1990	2010	1990	2010
1. Base Case									
Percent oil and gas	32	4	2	12	0	0	0	--	15
Percent coal	44	42	31	44	45	6	0	--	16
Percent nuclear	8	37	57	30	45	94	100	--	67
Percent other	16	17	10	14	11	0	0	--	2
3. Nuclear moratorium									
Percent oil and gas	32	4	2	12	0	0	0	--	
Percent coal	44	57	88	61	89	100	100	--	--
Percent nuclear	8	23	0	12	0	0	0	--	--
Percent other	16	17	9	14	11	0	0	--	--
4. Coal and shale limits									
Percent oil and gas	32	6	2	11	0	0	0	--	--
Percent coal	44	47	2	40	29	23	27	--	--
Percent nuclear	8	32	87	35	60	77	73	--	--
Percent other	16	17	9	13	11	0	0	--	--
6. Nuclear moratorium, coal and shale limits									
Percent oil and gas	32	4	6	12	0	0	0	--	--
Percent coal	44	60	71	69	77	100	100	--	--
Percent nuclear	8	18	0	5	5	0	0	--	--
Percent other	16	18	24	14	18	0	0	--	--

[a] For base-load generation only; the other models report percentage projections for all electricity generation (including intermediate and peak-load generation).
NOTE: Totals may not add due to rounding.

percentages of nuclear because the Nordhaus model includes only base-load generation, for which technologies like nuclear are most suitable. The other scenarios reinforce the earlier observations that for purposes of electricity generation, coal and nuclear power are close substitutes.

Table III.11 presents the MRG levelized cost assumptions under the Base Case for all electricity-generating technologies considered in the study. It demonstrates just how close a substitute coal is for nuclear power, and vice versa. As shown, over all the coal and nuclear technologies there is no more than a 5 percent spread in levelized costs. Solar-electric, however, is approximately twice as expensive, under Base-Case assumptions, as all other technologies.

The information in Table III.11 explains many of the results in each scenario. Coal and nuclear power are close substitutes for electricity generation because of the small cost difference between them. A similar observation holds with respect to the alternative nuclear reactors.

But having these cost data is not enough; otherwise, there would be no need for modeling. The critical factor in Table III.11 is the estimate of fuel costs, because it is this element of generation costs that is sensitive to particular policy and realization scenarios. If GNP growth is rapid, uranium and/or coal prices are likely to rise, altering the levelized costs of coal- and nuclear-based electricity. Similarly, if a moratorium is imposed on all ACR's and the FBR, uranium prices may rise sufficiently to induce a greater reliance on coal. The table, therefore, is a helpful guide to explaining the electricity-generation projections only for the Base Case and for those cases where relative fuel costs are not substantially altered.

Finally, it is of considerable interest to know what types of nuclear reactors the models project to be in use under different scenarios. Table III.12A presents these projections for the three groups of reactors considered by the MRG: the current LWR, all ACR's, and the FBR. As indicated in the note to the table, both ETA and SRI allow the LWR to recycle plutonium, whereas Nordhaus permits only advanced converters and breeders to perform recycling. This is an important definitional difference between the models and accounts for much of the intermodel differences in projections of nuclear reactor composition.

The ETA and Nordhaus projections are reported out to 2030 because, under Base-Case assumptions, the ACR and FBR technologies cannot be introduced until 2000, and are restricted in their rate of capacity growth thereafter. Looking only at 2010 projections, therefore, could be highly misleading, especially in scenarios where coal and shale oil production are limited or GNP growth is high.

With a few exceptions, the projections in Table III.12A indicate the introduction of an advanced converter technology by 2010 and, depending on the scenario, a heavy dependence on that technology by 2030. This is certainly the case for the Nordhaus and SRI projections, but less so for ETA, which allows LWR's to recycle plutonium and therefore reduces the need for more advanced reactors.

One interesting aspect of the projections is the small role projected for the FBR by ETA, Nordhaus, and SRI. Thus, only DESOM projects a more

TABLE III.11 Base-Case Levelized Cost Assumptions for Base-Load Electricity Plants at the 1975 General Price Level and at a 13 Percent Pretax Cost of Capital

Plant Type	Coal-fired (with Scrubbers)	Light Water Reactor (LWR)	Representative Advanced Converter (ACR), 0.82 Conversion Ratio	Fast Breeder Reactor (FBR)	Solar
Capital Costs ($/kW)	520	650	715	810	1,730
Unit costs (mills/kWh)					
Levelized capital costs[a]	12.2	15.2	16.7	19.0	40.5
Fuel costs (under the Base Case)	7.1[b]	4.9[c]	3.4[c]	2.1[c]	0
Operating and maintenance	3.0[d]	2.0	2.0	2.0	2.0
TOTAL	22.3	22.1	22.1	23.1	42.5

[a] Capital costs based on 13 percent/year discount rate, 30-year service life, and a 65 percent capacity factor. Therefore, levelized mills/kWh =

$$(0.1334/\text{year}) \{cap_i \div [0.65(8.65\ 10^3\ \text{hours/year})]\} = 0.0234\ cap_i.$$

[b] Environmental standards to be met by flue gas desulfurization $7.1/$10^3$ kWh = (coal costs at $0.75/$10^6$ BTU) (9,500 BTU/kWh) @ 36 percent overall efficiency.

[c] Nuclear fuel costs shown here are based upon uranium at $30 per pound and enrichment at $75 per separative work unit (SWU). No allowance is made for plutonium credits. Carrying charges on the fuel inventories are assumed to be 40 percent of the direct fuel costs (or approximately a 13 percent annual carrying charge on a 3-year inventory).

[d] Includes cost of sludge disposal.

NOTE: All calculations presented in this table were performed by Alan Manne.

TABLE III.12A Nuclear Fission Energy Production, by Type of Reactor, in 2010 and 2030, Under Different Policy and Realization Scenarios[a] (Units: Quads of Primary Energy Equivalents)

Policies and Realizations: Base Case, Modified as Stated (Scenario Number Indicated at Left)	DESOM	ETA		Nordhaus		SRI
	2010	2010	2030	2010	2030	2010
1. Base Case						
Total nuclear	*35.4*	*31.2*	*29.1*	*41.8*	*68.7*	*52.8*
LWR	28.3	31.2	35.8	32.7	5.2	17.5
ACR	0.3	0.0	3.3	9.1	63.4	33.5
FBR	6.8	0.0	0.0	0.0	0.0	1.8
2. Moratorium on all ACR's and the FBR						
Total nuclear	*18.4*	*30.9*	*38.2*	*41.8*	*37.2*	--
4. Coal and shale limits						
Total nuclear	*57.4*	*41.0*	*80.7*	*30.3*	*72.7*	--
LWR	48.6	40.7	21.7	25.8	6.7	--
ACR	1.8	0.3	58.7	4.5	66.1	--
FBR	6.8	0.0	0.4	0.0	0.0	
5. Moratorium on all ACR's and the FBR and coal and shale limits						
Total nuclear	*55.7*	*39.8*	*71.5*	*28.1*	*66.5*	--
7. High GNP						
Total nuclear	--	*55.0*	*89.8*	*55.1*	*147.8*	*87.1*
LWR	--	48.2	16.1	41.9	2.1	21.9
ACR	--	6.8	59.4	13.2	129.3	57.9
FBR	--	0.0	14.3	0.0	16.4	7.4
8. Low GNP						
Total nuclear	--	*15.3*	*13.0*	*26.0*	*27.2*	*30.3*
LWR	--	15.3	13.0	26.0	27.2	13.4
ACR	--	0.0	0.0	0.0	0.0	16.5
FBR	--	0.0	0.0	0.0	0.0	0.4

[a]ETA and SRI include LWR's with plutonium recycle in the LWR category; Nordhaus excludes these reactors from the category of LWR's but includes them as an ACR.

significant role in 2010 for the FBR than an ACR in the Base Case.* A possible explanation for this is that DESOM adds the electricity required for uranium enrichment to total electric generation, although this feature alone would not appear to favor the introduction of the FBR in either ETA or Nordhaus. In contrast to DESOM, ETA and Nordhaus project substantial use of the FBR only in the high-GNP scenarios. In scenarios of low-GNP growth or partial or total nuclear moratorium, neither an advanced converter nor the FBR is introduced.

Because of the importance of the projections of reactor mixes in the CONAES study, the MRG examined the sensitivity of these projections to two important variables: (1) uranium availability and (2) the investment and pricing discount rate. Because these sensitivity tests were conducted at various times different from the majority of the MRG work, however, they are based on a slightly different set of assumptions than those used in the other MRG simulations. In the DESOM and Nordhaus uranium sensitivity tests, these assumptions include:

1. Synthetic fuels available at an average cost of $2.55/10^6 BTU (MRG Base Case--$3.70/10^6 BTU)
2. Non-price-induced conservation is assumed to have reduced energy consumption by 15 percent in 2010 (MRG Base Case--10 percent)
3. The ACR and FBR are assumed to be commercially available by 1990 and 1995, respectively (MRG Base Case--2000 for both).

For ETA, on the other hand, the uranium sensitivity test assumed:

1. No non-price-induced conservation
2. GNP growth of 3.06 percent per annum over the 1975-2010 period (MRG Base Case--3.19 percent)
3. No ACR will be commercially available.

Clearly, because of these differences in assumptions, the uranium sensitivity test is more instructive within than between models. Even within models, the sensitivity test with respect to the Base Case is not a clear-cut one because of the differences in assumptions between Base-Case and non-Base-Case sensitivity scenarios just mentioned. Nevertheless, these results are shown because for the question of reactor mix alone, the effects of the differences in assumptions should be small in comparison with the influence of uranium availability.

Table III.12B presents the reactor mix projections under different uranium availability realizations. Apart from the MRG Base Case, the table indicates the sensitivity of both total nuclear energy production and reactor mixes to an "intermediate" availability of 3.7 million tons at prices up to $45/lb U_3O_8, but none at higher prices, and to the MRG low uranium availability of no more than 2.25 million tons at prices up to $225/lb.

*The DESOM projection of nuclear production by advanced reactors (ACR and FBR) in the Base Case and the moderate coal and shale limits case is virtually identical to its projection in the scenario of low coal and shale limits.

TABLE III.12B Nuclear Fission Energy Production, by Type of Reactor, in 2010 and 2030, Under Different Uranium-Availability Assumptions (Units: Quads of Primary Energy Equivalents)

	DESOM	ETA		Nordhaus	
Uranium availabilities	2010	2010	2030	2010	2030
1. MRG Base Case (3.7 million tons at minimum acceptable prices up to \$45/lb U_3O_8; 12.57 million tons total)					
Total nuclear	*35.1*	*31.2*	*39.1*	*41.8*	*68.7*
LWR	28.3	31.2	35.8	32.7	5.2
ACR	0.3	0.0	3.3	9.1	63.4
FBR	6.8	0.0	0.0	0.0	0.0
2. Intermediate (3.7 million tons at minimum acceptable prices up to \$45/lb U_3O_8; none at higher prices)					
Total nuclear	--	*34.1*	*45.8*	--	--
LWR	--	33.8	0.0	--	--
ACR	--	0.0	4.6	--	--
FBR	--	0.3	41.2	--	--
3. Low (1.6 million tons at minimum acceptable prices up to \$45/lb U_3O_8; 2.25 million tons total)					
Total nuclear	*31.8*	*36.5*	*76.5*	*31.6*	*55.5*
LWR	26.2	31.2	10.4	20.8	0.0
ACR	0.3	0.0	0.0	10.8	55.5
FBR	5.3	5.3	66.1	0.0	0.0

NOTE: See text for explanation of slightly different assumptions underlying non-Base-Case results.

The clearest implications are provided by ETA, which shows increasing reliance on advanced reactors, and on the breeder in particular, as uranium availability falls. A somewhat surprising outcome, however, is the ETA projection of production by LWR's, which falls off rapidly between 2010 and 2030 in the non-Base-Case uranium scenarios. This outcome can be understood as a projection that no new LWR facilities would be added after 2010 and that all new reactor construction would center on the ACR and/or the FBR. The LWR's in existence in 2010 would then be retired after their normal lifetimes expire.

The DESOM and Nordhaus results are more ambiguous, largely because of the difference between synthetic fuels prices assumed in the MRG Base

Case and in the low-uranium-availability scenario. Thus, a lower synthetic fuels price in the last scenario reduces the relative need for electricity in general (in the industrial, commercial, and residential sectors, but less so in transportation), especially from advanced reactors. Nevertheless, both DESOM and Nordhaus project increasing reliance on advanced reactors as uranium availability declines--DESOM favoring the breeder, Nordhaus projecting the use only of advanced converters.

The implications of low uranium availability for advanced reactor production are clearer when viewed with the simultaneous imposition of low coal and shale limits. Thus, in this "scarce resource" case, DESOM projects (below) greater reliance in 2010 on nuclear power, in general, and advanced reactors, in particular:

	Base Case	Low Uranium Availability	Low Uranium Availability and Low Coal and Shale Limits
Total nuclear production (in Quads)	*35.4*	*31.8*	*43.8*
LWR	28.3	26.2	35.2
ACR	0.3	0.3	1.8
FBR	6.8	5.3	6.8

Viewed as a whole, therefore, the sensitivity test illustrates that the uranium availability assumption can be a critical force in projecting future reactor production and mix.*

In the second sensitivity test, ETA simulated an earlier version of the Base Case under two different discount rate assumptions. One set of assumptions represents the Base-Case values of 6 percent and 13 percent for consumers' and producers' decisions, respectively. The other introduces a 10 percent rate to be applied only to all public utility investment and pricing decisions, on the basis that public utilities have historically been allowed lower rates of return than manufacturing enterprises by state regulatory commissions. The earlier version of the Base Case used to test discount rate sensitivity differs from the final version in these respects:

1. The non-price-induced conservation effect is assumed to be 15 percent (MRG Base Case--10 percent).
2. Average cost of synthetic fuels is assumed to be $2.55/$10^6$ BTU (MRG Base Case--$3.70/$10^6$ BTU).
3. Uranium resources are assumed to be represented by the "high" column in Table II.1.
4. Commercial availability of the ACR and FBR assumed to be 1990 and 1995, respectively (MRG Base Case--$3.70/$10^6$ BTU).

*This result is confirmed in Section III.8, where the net benefit of the FBR is estimated under different uranium availability assumptions.

Although these differences in assumptions can have a substantial impact on the absolute levels of certain variables, it is unlikely that they would materially affect the projections of the differences in outcomes generated by variations in the discount rate.

The outcomes reported in Table III.13 indicate that a lower rate of discount for public utility pricing and investment decisions would lead to slightly increased levels of total energy consumption and substantial increases in FBR production relative to the Base Case. As the footnote to the table indicates, the ETA simulations do not incorporate the ACR. Nevertheless, the sensitivity test reveals, at the very least, that the energy production by advanced reactors taken together is likely to be highly sensitive to the discount rate. Other outcomes of the sensitivity test of interest (not reported numerically) include the findings that total electricity generation increases approximately 6 percent, and electricity prices fall nearly 15 percent by 2010 under the 10 percent discount rate assumption. Oil and gas prices, in the long run, remain largely the same in the presence of either discount rate.

Finally, after all the attention given to the question of reactor mixes, the issue of future uranium consumption merits attention of its own. Figures III.6A and III.6B illustrate projections by DESOM and ETA of cumulative uranium consumption for up to eight different scenarios. The projections are reported out to 2020 for DESOM and to 2030 for ETA because of the continued growth in demand for uranium arising from nuclear plants now in use and those that may be constructed in the future.

Except for scenarios of coal and shale oil limits or high GNP (nos. 5 and 7 in the chart), cumulative uranium consumption is not projected by ETA to have exceeded 3.7 million tons by 2030. DESOM projects cumulative uranium consumption by 2020 to exceed 3.7 million tons only in the two cases where coal and shale limits are imposed. The 3.7 million figure represents the quantity of uranium resources assumed, in the Base Case, to be available at a minimum acceptable price of less than or

TABLE III.13 ETA Projections, for 2010 and 2030, of Selected Output Variables Under Two Different Rates of Discount Applicable to Public Utility Pricing and Investment Decisions

Output Variables	Discount Rate: 13 Percent	Discount Rate: 10 Percent
Total energy consumption (Quads)		
2010	171.5	177.7
2030	296.2	300.3
FBR energy production (Quads)[a]		
2010	0.0	5.3
2030	0.0	13.7

[a]These simulations assume that an ACR is unavailable.

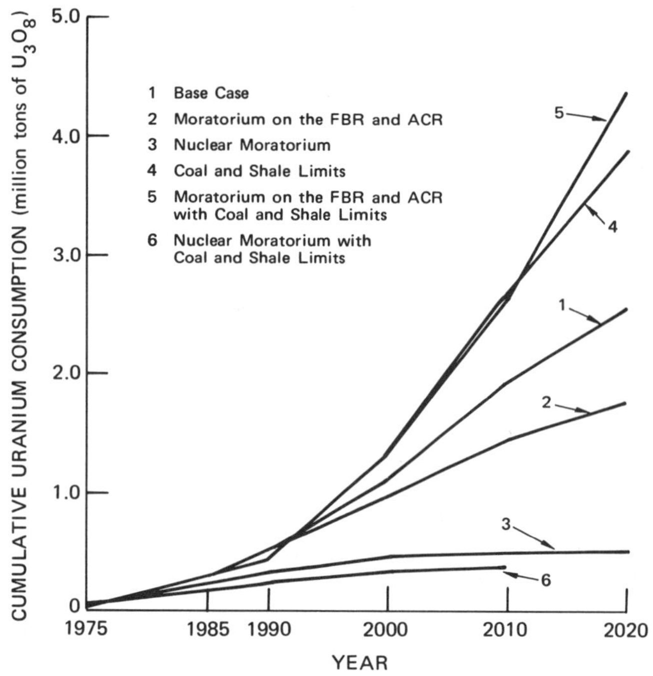

FIGURE III.6A DESOM projections of cumulative uranium consumption, 1975-2020, under alternative scenarios.

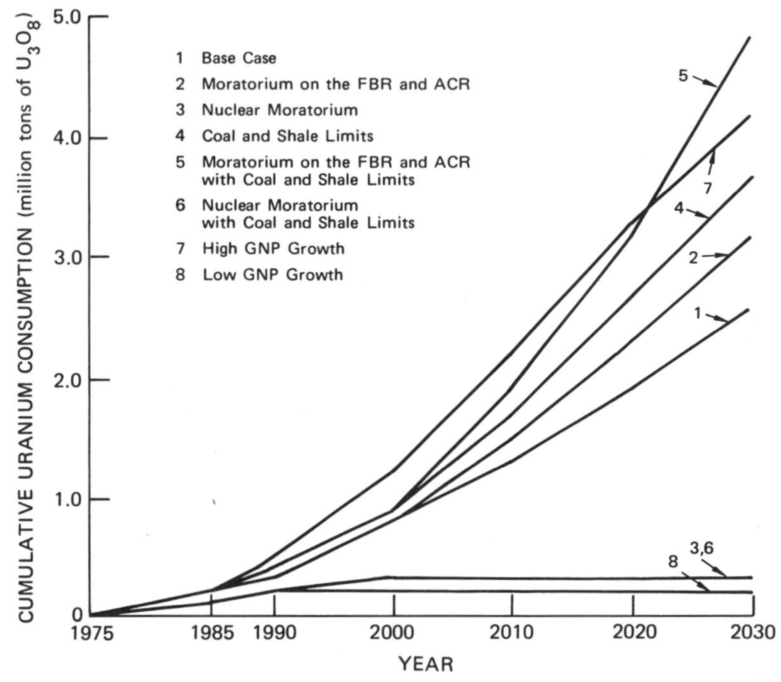

FIGURE III.6B ETA projections of cumulative uranium consumption, 1975-2030, under alternative scenarios.

equal to $45/lb (see Table II.1). If higher-priced uranium is available (under the MRG Base-Case assumption, an additional 8.87 million tons is available at prices between $45/lb and $225/lb), then under all of the scenarios uranium consumption is projected, at least through 2030, to fall safely within available supplies.

The DESOM and ETA projections are substantially confirmed by the Nordhaus results. Although the Nordhaus model does not specifically report cumulative consumption, such figures can be inferred from the projections of nuclear production, by reactor type.

III.6 DEMAND MIX PROJECTIONS: THE ROLE OF ELECTRICITY

In the face of dwindling low-cost oil and gas supplies, many predict an increasing role for electricity. The potential for the substitution of more abundant resources for oil and gas as inputs to electricity generation lends credibility to such predictions. So does the potential for the substitution of electricity for fossil fuels in many end uses. This section discusses the model projections for both the absolute growth in electricity generation and the role electricity plays in the various scenarios.

Tables III.14 and III.15 show the electric and nonelectric energy consumption projections made by the models for the various policy and realization scenarios, respectively. Electricity consumption is measured at the point where electricity leaves the generation plant (busbar). Nonelectric energy consumption is measured at the wellhead or port of entry for oil and gas not used for electricity generation, at the mine mouth for coal that is not used for production of synthetic fuels or electricity generation, and for synthetic fuels at the output of the synthetic fuel plant. The total consumption figures reported here, therefore, are equal to total domestic energy production plus imports (see Table III.3 for these totals) less the losses in electricity generation and in the manufacture of synthetic fuels.*

The projections in Table III.14 indicate that electricity growth is largely independent of policy choice. Only in the case of the simultaneous imposition of a nuclear moratorium and of limits on coal and shale oil production do each of the models indicate at least a moderate reduction in electricity consumption below Base-Case levels. These results are consistent with the finding in Section III.5 that, for electricity generation, coal and nuclear power are close substitutes, suggesting that restrictions on either should not materially affect electricity consumption. When restrictions are placed on both fuels at the same time, however (scenario 6 in Table III.14), the lack of substitutes

*These consumption totals exceed the total amount energy consumers actually receive by the average losses attributable to refining, transportation, and transmission, etc., or by about 5-6 percent of total energy actually consumed.

TABLE III.14 Electric and Nonelectric Energy Consumption Under Different Policy Scenarios[a] (Units: Quads of Primary Energy Equivalents)

Policies: Base Case, Modified as Stated (Scenario Number Indicated at Left)	ETA			Nordhaus			DESOM		
	Electric	Nonelectric	Total	Electric	Nonelectric	Total	Electric	Nonelectric	Total
1. Base Case									
1990	11.3	76.0	87.3	9.6	67.9	77.5	10.6	83.6	94.2
2000	16.4	80.9	97.3	11.8	73.4	85.2	15.6	105.8	121.4
2010	23.9	89.6	113.5	14.1	84.9	99.0	21.1	136.3	157.4
2. Moratorium on all ACR's and the FBR									
1990	11.3	76.0	87.3	9.6	67.9	77.5	10.6	83.7	94.3
2000	16.4	80.9	97.3	11.8	73.4	85.2	15.6	103.8	119.4
2010	23.6	89.5	113.1	14.1	84.9	99.0	21.1	137.6	158.7
3. Nuclear moratorium									
1990	10.9	76.4	87.3	8.9	64.0	72.9	10.6	82.5	93.1
2000	16.0	80.7	96.7	11.0	70.6	81.6	15.6	105.4	121.0
2010	23.2	89.4	112.6	13.6	81.6	95.2	20.5	136.7	157.2
4. Coal and shale oil limits									
1990	11.6	74.5	86.1	9.2	66.7	75.9	10.6	82.9	93.5
2000	15.7	81.7	97.4	11.8	73.9	85.7	16.7	104.7	121.4
2010	23.5	86.6	110.1	14.1	85.1	99.2	22.6	122.6	145.2
5. Moratorium on all ACR's and the FBR, coal and shale oil limits									
1900	11.6	74.5	86.1	8.9	66.1	75.0	10.6	83.0	93.6
2000	15.7	81.7	97.4	11.5	73.1	84.6	15.6	105.0	120.6
2010	23.2	86.8	110.0	14.1	84.5	98.6	22.9	121.4	144.3
6. Nuclear moratorium, coal and shale oil limits									
1990	11.3	74.7	86.0	8.1	56.9	65.0	10.2	83.7	93.9
2000	15.0	81.5	96.5	10.2	64.0	74.2	14.4	103.3	117.7
2010	16.7	95.2	111.9	12.3	80.4	92.7	18.5	126.8	145.3

[a] Electric energy consumption measured at the busbar; nonelectric energy consumption measured net of synfuel conversion losses, but before refining, distribution, and transmission losses.

TABLE III.15 Electric and Nonelectric Energy Consumption Under Different Realization and Policy Scenarios (Unit: Quads of Primary Energy Equivalents)

Policies: Base Case Modified as Stated (Scenario Number Indicated at Left)	ETA			Nordhaus		
	Electric	Non-Electric	Total	Electric	Non-Electric	Total
1. Base Case						
1990	11.3	76.0	87.3	9.6	67.9	77.5
2000	16.4	80.9	97.3	11.8	73.4	85.2
2010	23.9	89.6	113.5	14.1	84.9	99.0
7. High GNP growth						
1990	15.0	84.8	99.8	11.4	79.7	91.1
2000	27.7	82.2	109.9	16.1	95.5	111.6
2010	40.3	120.6	160.9	22.5	122.0	144.5
8. Low GNP growth						
1990	9.6	69.4	79.0	7.9	58.2	66.1
2000	10.9	67.4	78.3	8.5	55.1	63.6
2010	13.0	61.0	74.0	8.8	56.2	65.0
9. Low GNP growth, nuclear moratorium, coal and shale oil limits, high-cost nonelectric AES						
1990	9.2	68.6	77.8	7.1	51.0	58.1
2000	11.6	64.6	76.2	8.1	53.7	61.8
2010	12.3	61.3	73.6	8.8	56.9	65.7

[a]Electric energy consumption measured at the busbar; nonelectric energy consumption measured net of synfuel conversion losses, but before refining, distribution, and transmission losses.

at comparable costs forces increases in electricity prices and, therefore, cutbacks in consumption.*

For the Base Case, in particular, electricity consumption between 1975 and 2010 is projected by the models to grow at an annual rate between 2.1 percent (Nordhaus) and 3.7 percent (ETA). These projections should be contrasted with recent electric utility forecasts of growth between 1975 and 1990 of between 5.3 percent and 5.8 percent per annum.† Even with allowances for their shorter time horizon, the industry forecasts are higher than the highest projection made by any model here for any scenario.

There are two major reasons why the electricity growth-rate projections in this study are substantially below the 7 percent rate of growth that has been experienced since 1945. First, and most important, the MRG projects the price of electricity to increase approximately 1 percent per annum (see Section III.7) over the next 35 years in contrast to the approximate 4 percent per annum *decline* in the real price of electricity over the 1940-1975 period. The arrest in the historical decline of electricity prices, coupled with the price sensitivity of electricity demand, should contribute to slower growth rates of electricity demand than have previously occurred, unless new technologies of electricity use have a decisive impact.

Second, within a fairly wide range of prices, the models generally do not allow substantial substitution of electric for nonelectric energy. This feature is to be distinguished from the price responsiveness of *total* primary energy demands, which is reflected in all of the models except DESOM. As a result, the models do not project large-scale substitutions of electricity for nonelectric energy, even though nonelectric energy prices are projected to rise relative to the price of electricity over the 1975-2010 period.‡

Within this broad range of agreement about the general direction of electricity growth, the models still exhibit widely varying growth-rate

*As Table III.11 indicates, under MRG cost assumptions, the closest alternative to either coal or nuclear power for electricity generation is solar power, at approximately twice the cost. Except for ETA, however, the models do not project a conversion to solar electricity under the policy scenario of nuclear, coal, and shale oil restrictions, since there still is sufficient coal for electricity generation. Nevertheless, the price of coal is projected to increase under these policy conditions, which explains the projected increase in electricity prices. For all price projections, see Section III.7.

†See, e.g., *Economic Growth in the Future,* Edison Electric Institute, 1976.

‡In non-CONAES simulations with ETA-MACRO, an integrated energy sector/macroeconomic growth model developed by Alan Manne, electricity demand growth rates were found to be sensitive to the degree of substitution in production between electric and nonelectric energy. See Alan S. Manne, "ETA-MACRO: A Model of Energy-Economy Interactions," presented at ORSA/TIMS meeting, San Francisco, May 9-11, 1977.

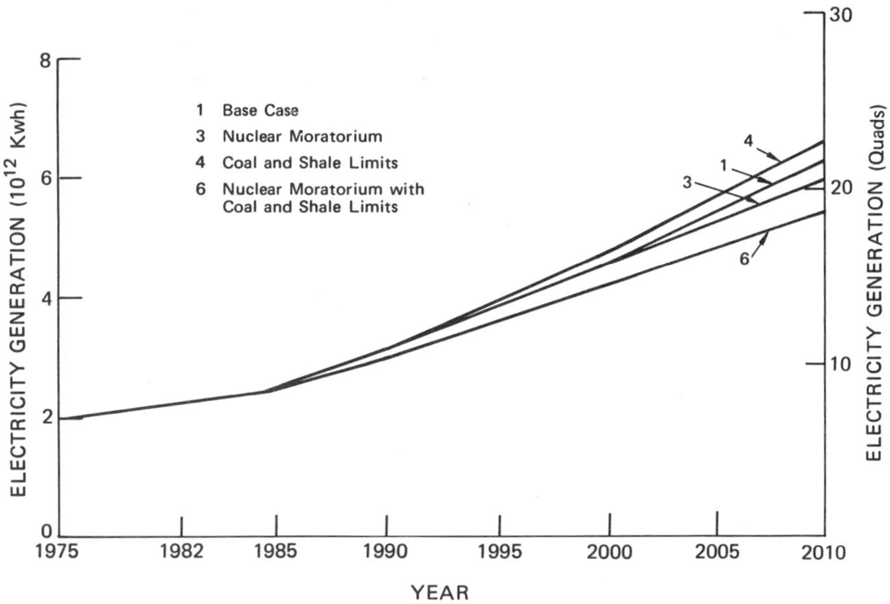

FIGURE III.7A DESOM electricity-generation projections.

projections. Both Table III.14 and Figures III.7A and III.7B indicate that the projections of electricity growth by Nordhaus are significantly lower than those of DESOM and ETA. In part, this is due to the lower nonelectric energy prices projected by Nordhaus than by the other two models. Equally important, the three models differ in their treatment of the responsiveness of electric energy demand to changes in the price of nonelectric energy. In the context of the above discussion, the models differ in the degree to which they allow electricity to substitute for nonelectric forms of energy. Demand for electricity in the DESOM model, for example, is determined by the degree of interfuel substitution that minimizes the attained value of the cost function. The cross-elasticity (measuring the quantity response of one good to price changes in another good) with respect to the price of nonelectric energy, therefore, is highly discontinuous in DESOM, jumping between zero and infinity. A similar result occurs over the price ranges in the MRG scenarios for the Nordhaus model,* although the DESOM projections remain above the Nordhaus projections because of the built-in assumption in DESOM that end-use demands for energy services do not respond to price changes. A constant nonzero cross-elasticity is incorporated in ETA, however, which explains why its electricity growth projections are substantially above those made by Nordhaus.

While projections in the growth of electricity consumption are highly useful, it is just as important to examine the role of electricity

*See Appendix C for a description of the discontinuous nature of the cross-elasticities for fuel demands embodied in the Nordhaus model.

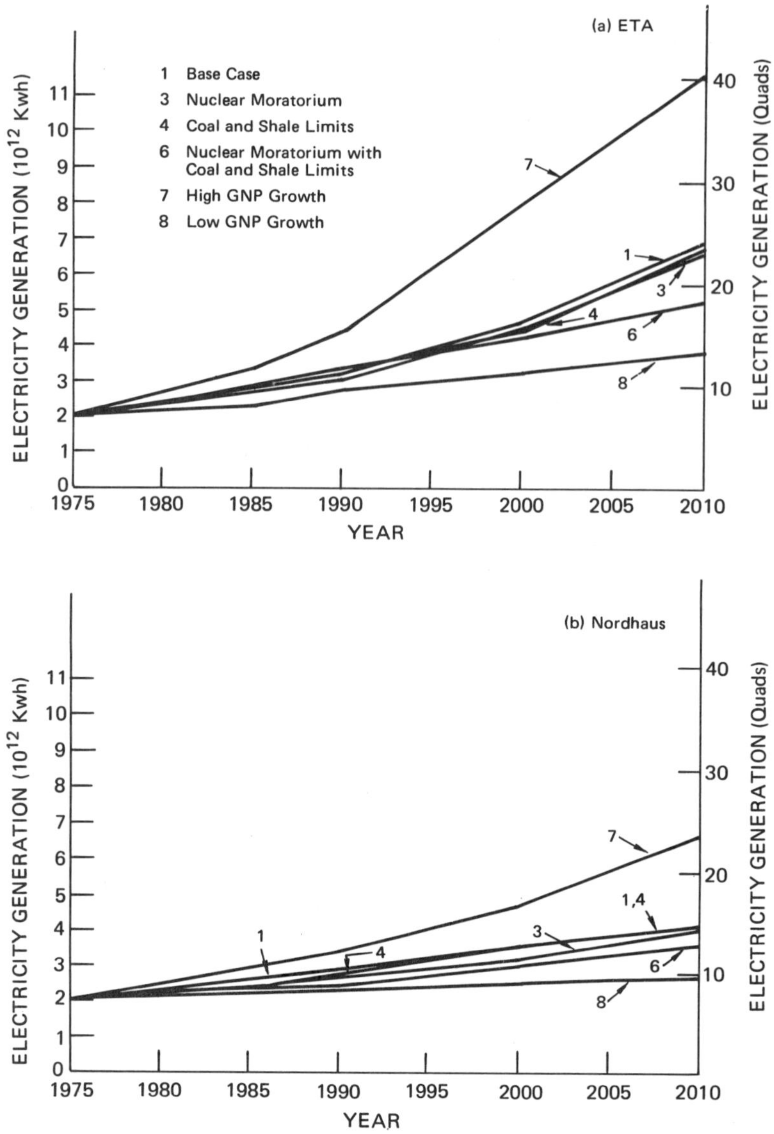

FIGURE III.7B ETA and Nordhaus electricity-generation projections.

relative to other forms of energy. Consequently, Figures III.8A and III.8B illustrate projections of the ratio of electricity to total consumption under alternative scenario assumptions.

In the DESOM and Nordhaus results, the role of electricity is only slightly dependent on the scenario chosen. Both models project the share of electricity in total consumption to rise from approximately 10.6 percent in 1975 to between 12 percent and 15.6 percent in 2010, depending on the scenario. This result is consistent with the way each model treats the cross-elasticity of demand for electricity. ETA, however, which incorporates a nonzero cross-elasticity of electricity demand, projects the share of electricity consumption to range between 15 percent and 25 percent in 2010. The high GNP growth scenario in the ETA simulations requires electricity to play its biggest role; the projected

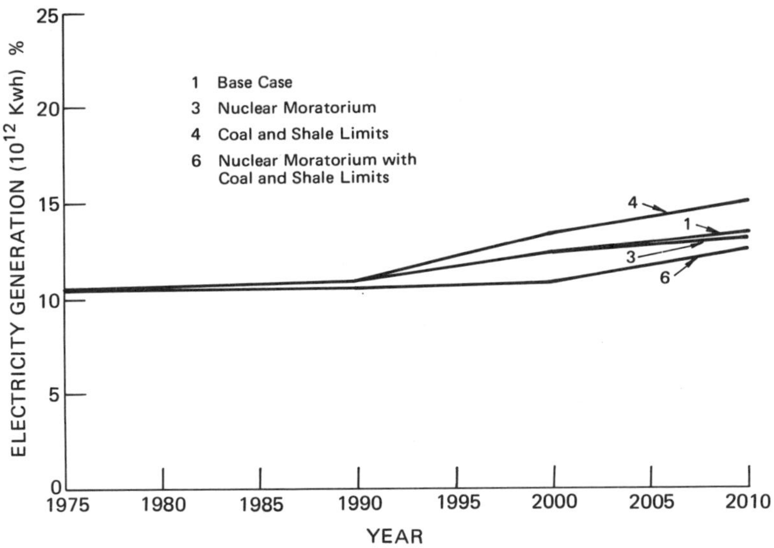

FIGURE III.8A DESOM projections of electricity consumption as a percentage of total energy consumption.

share of electricity in total consumption is considerably lower in the other scenarios. Nevertheless, the projected percentage of electricity in the final demand mix is greater in almost any ETA scenario projection than in any made by DESOM and Nordhaus.

Despite their differences, all models agree on the relative effects on electricity shares that may flow from the adoption of the policies and from the GNP realizations considered by the MRG. Restrictions on either nuclear power or coal are projected to have negligible effects on the electricity share. GNP realizations, on the other hand, are projected to have much greater effects, electricity occupying greater shares of total consumption with higher GNP growth rates. There appears to be a ceiling on the electricity share, however, in the ETA simulation of the high-GNP scenario, attributable to the availability of synthetic fuels at competitive prices. Finally, the role of electricity is significantly reduced according to all models in 2010 if both a nuclear moratorium and coal and shale limits are imposed. In fact, if these policies are adopted, electricity's relative role is projected by ETA and Nordhaus to be smaller than in the low-GNP-growth scenario, since nuclear and coal restrictions would have a greater effect on electricity demand, in particular, than the realization of low GNP.

Finally, in its earlier work, the MRG specified one scenario that serves as an "electrification sensitivity test." This scenario, based on somewhat different Base-Case central values than those finally agreed upon, envisioned that a synthetic fuels technology would be available in 1995 at a cost of \$4.50/10^6 BTU, somewhat higher than the current Base-Case value of \$3.70/10^6 BTU. Under this cost assumption, Nordhaus projected no change (relative to an earlier version of the Base Case) in the absolute quantity of electricity consumed in 2010, and almost no change in the relative share of electricity as a percentage of total energy consumption. DESOM projected a slight increase in 2010 relative to an earlier version of the Base Case in both the absolute quantity and relative share of electricity consumption.

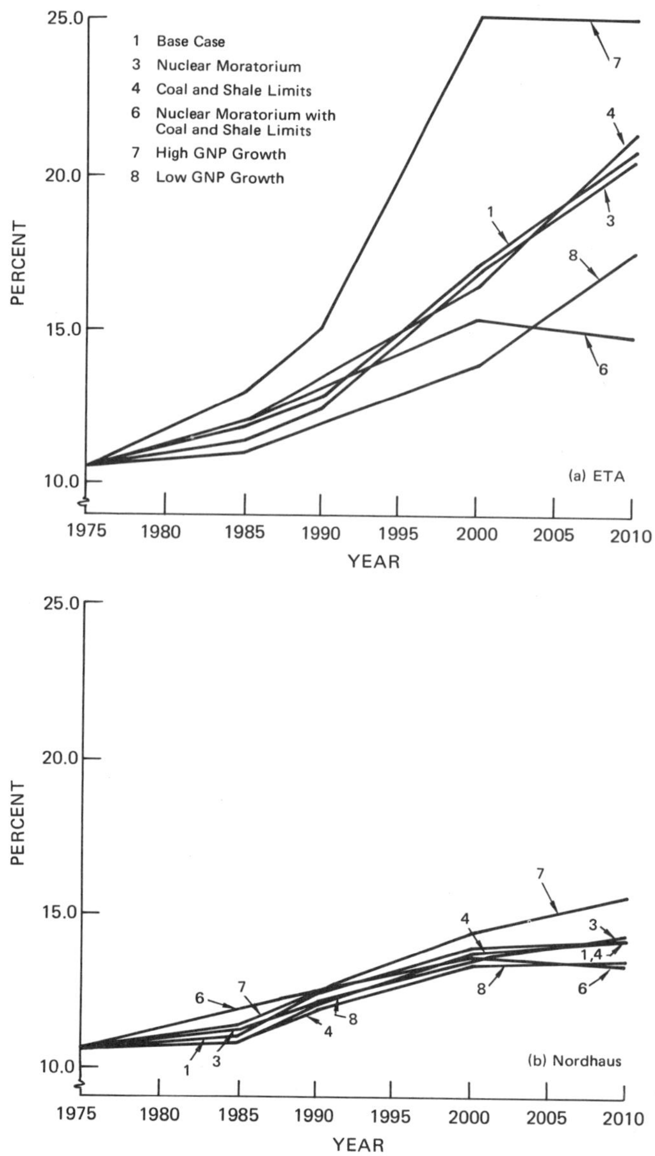

FIGURE III.8B ETA and Nordhaus projections of electricity consumption as a percentage of total energy consumption.

The striking aspect of both of these results is that they project relatively no change in the share of electricity in response to a wide range of synthetic fuels costs. In particular, since the earlier version of the Base Case assumed synthetic fuels to cost an average of $2.55/10^6$ BTU, the assumption of $4.50/10^6$ BTU for purposes of sensitivity analysis represents close to a doubling in cost. Nevertheless, both Nordhaus and DESOM project little change in electrification.

These results must be tempered, however, by the fact that electricity may be less susceptible to supply interruptions than oil, natural gas, or synthetic fuels. Current conversions to electricity appear to be motivated largely by this concern for reliability. Moreover, neither DESOM nor Nordhaus incorporates cross-elasticities of demand between electric and nonelectric sources of energy.* Interfuel substitution, instead, is modeled through cost minimization. Taken together, these qualifications suggest that the degree of electrification may be more sensitive to synthetic fuels prices than the bare DESOM and Nordhaus results would appear to indicate.

III.7 LONG-RUN ENERGY PRICE TRENDS

This section presents the long-run real (constant-dollar) price trends projected by the models for the alternative policy and realization scenarios. Tables III.16 and III.17 show the long-run price projections for selected primary resources for the policy and realization scenarios, respectively, while Tables III.18 and III.19A show the corresponding price projections for electricity. The projected energy prices show a tendency to increase over time, but the price increases projected are by no means alarming. In particular, no price projected by any model for any scenario increases at an average annual rate in excess of 3 percent between now and the year 2010.

In interpreting the results listed in Tables III.16 through III.19B, several assumptions embodied in the models should be kept in mind. First, there are differences in the pricing concepts used in the alternative models. One difference is that ETA and Nordhaus report prices for electricity at the busbar (at the point where electricity leaves the generation plant), for oil and natural gas at the wellhead, and for coal at the mine mouth, whereas SRI and DESOM report prices at the point of delivery. These differences are indicated in Tables III.18 and III.19A. Because of costs and markups between these two stages, prices reported by SRI and DESOM should always exceed those reported by ETA and Nordhaus.

Second, the prices of all primary or refined forms of energy are to be thought of as equilibrium prices in a national energy market. This is so regardless of whether the equilibrium is explicitly described by the model (as in SRI) or (as explained in Section I.4) is simulated by an optimization that generates implicit equilibrium prices as "shadow prices." The following stipulations follow in one way or another from this concept.

Third, in all the models it is assumed that oil and gas prices will be deregulated in the near future. Since oil and gas prices are currently regulated, albeit less restrictively than in the recent past, the projected oil and gas prices in the short run look high in comparison with prices currently prevailing. The price of imported oil, however, is assumed to remain at $13/bbl in real terms over the entire projection period (see Table II.1). This Base-Case assumption steers a middle ground between the view on the one hand that OPEC may collapse and the opposite position, which foresees further substantial increases in the real price of imported oil.

*This feature is contained in ETA, but due to time constraints, the "electrification sensitivity test" was not simulated on the ETA model.

TABLE III.16 Prices of Selected Primary Energy Resources Projected by the Long-Run Models Under Different Policy Scenarios[a]

Policies: Base Case Modified as Stated (Scenario Number Indicated at Left)	ETA			Nordhaus		DESOM	
	Oil ($/bbl)	Coal ($/$10^6$ BTU)	Uranium ($/lb)	Oil ($/bbl)	Coal ($/$10^6$ BTU)	Oil ($/bbl)	Coal ($/$10^6$ BTU)
Actual, 1975	9.00	0.75	30.0	9.00	0.75	9.00	0.75
Projected:							
1. Base Case							
1990	14.50	0.75	32.0	14.80	0.75	14.39	1.17
2000	18.56	0.75	33.6	15.95	0.75	17.54	1.17
2010	23.78	0.75	36.5	17.40	0.75	23.22	1.17
2. Moratorium on all ACR's and the FBR							
1990	14.50	0.75	33.6	14.80	0.75	14.42	1.17
2000	18.56	0.75	36.4	15.96	0.75	17.57	1.17
2010	24.36	0.75	41.5	17.43	0.75	23.11	1.17
3. Nuclear moratorium							
1990	15.08	0.75	30.0	14.72	1.31	14.77	1.17
2000	19.14	0.75	30.0	15.81	0.75	18.89	1.17
2010	24.94	0.75	30.0	17.48	0.75	16.89	1.17
4. Coal and shale oil limits							
1990	15.66	0.75	34.9	25.59	1.22	15.51	1.17
2000	18.56	0.75	38.8	19.61	0.60	21.58	1.17
2010	29.00	1.50	45.7	19.81	0.65	29.25	3.02
5. Moratorium on all ACR's and the FBR, coal and shale oil limits							
1990	15.66	0.75	39.9	15.08	1.24	15.51	1.17
2000	18.56	0.75	47.7	16.46	0.61	21.59	1.17
2010	29.00	1.50	61.7	18.44	0.66	28.31	2.86
6. Nuclear moratorium, coal and shale oil limits							
1990	16.24	0.75	30.0	16.07	2.07	15.45	1.32
2000	19.72	1.00	30.0	17.22	1.54	21.35	1.97
2010	29.00	2.70	30.0	19.60	1.47	25.25	4.60

[a]Price of oil at the wellhead, price of coal delivered to industry, price of uranium after milling; all prices in 1975 dollars.

TABLE III.17 Prices of Primary Energy Resources Projected by the Long-Run Models Under Different Realization and Policy Scenarios

Realizations: Base Case, Modified as Stated (Scenario Number Indicated at Left)	ETA Oil ($/bbl)	ETA Coal ($/10^6 BTU)	ETA Uranium ($/lb)	Nordhaus Oil ($/bbl)	Nordhaus Coal ($/10^6 BTU)
Actual, 1975	9.00	0.75	30.0	9.00	0.75
Projected:					
1. Base Case					
1990	14.50	0.75	32.0	14.80	0.75
2000	18.56	0.75	33.6	15.95	0.75
2010	23.78	0.75	36.5	17.40	0.75
7. High GNP growth					
1990	17.40	0.75	37.4	15.74	0.75
2000	29.00	0.75	43.2	17.37	0.75
2010	29.00	0.75	53.7	20.15	0.76
8. Low GNP growth					
1990	13.34	0.75	30.0	14.06	0.75
2000	15.08	0.75	30.0	14.63	0.75
2010	21.46	0.75	30.0	15.38	0.75
9. Low GNP growth, nuclear moratorium, coal and shale oil limits, high-cost nonelectric AES					
1990	15.08	0.75	30.0	14.95	1.66
2000	17.40	0.75	30.0	15.01	0.86
2010	22.04	0.90	30.0	16.05	0.51

[a]Price of oil at the wellhead, price of coal delivered to industry, price of uranium after milling; all prices in 1975 dollars.

Fourth, a related set of issues arises in connection with the projections of electricity prices, which are determined by the assumed levels of capital costs and projected resource prices. Under current regulatory practice, electricity prices are set equal to the average *historical* cost of generating, transmitting, and delivering electricity. In the models used by the MRG, however, the costs of all electric facilities are set at *current replacement* costs (see Table II.1), which are normally higher than historical costs. Consequently, price projections over the short run, made on the basis of MRG cost assumptions, are likely to overstate actual electricity prices. As existing facilities are replaced by new higher-cost units, however, actual electricity prices are

TABLE III.18 Prices of Electricity Projected by the Long-Run Models Under Different Policy Scenarios (All Prices in 1975 Dollars)

Policies: Base Case Modified as Stated (Scenario Number Indicated at Left)	Busbar (Shadow) Price ($/1,000 kWh)		Delivered Industrial Price ($/1,000 kWh)	Delivered Consumer's (Shadow) Price ($/1,000 kWh)
	ETA	Nordhaus	SRI	DESOM
1975 Price: Busbar--$15/1,000 kWh[a] Delivered--$23/1,000 kWh				
Projected:				
1. Base Case				
1990	21.6	20.4	29.2	35.7
2000	21.6	19.7	28.4	36.1
2010	21.6	20.1	27.3	36.9
2. Moratorium on all ACR's and the FBR				
1990	21.8	20.4	--	35.7
2000	21.8	19.7	--	36.2
2010	22.0	20.1	--	36.9
3. Nuclear moratorium				
1990	23.0	26.4	--	35.7
2000	23.0	21.1	--	35.0
2010	23.0	21.1	--	36.6
4. Coal and shale oil limits				
1990	21.3	25.6	--	32.2
2000	22.8	19.6	--	38.8
2010	25.2	19.8	--	39.9
5. Moratorium on all ACR's and the FBR, coal and shale oil limits				
1990	21.7	25.7	--	30.9
2000	23.2	19.7	--	43.9
2010	25.7	20.2	--	37.6
6. Nuclear moratorium, coal and shale oil limits				
1990	23.0	33.6	--	37.5
2000	25.3	28.6	--	44.6
2010	41.1	27.9	--	84.9

[a]Calculated by Alan Manne.

TABLE III.19A Prices of Electricity Projected by the Long-Run Models Under Different Realization Scenarios (At the Busbar in 1975 Dollars)

Realizations: Base Case Modified as Stated (Scenario Number Indicated at Left)	Busbar (Shadow) Price ($/1,000 kWh)	
	ETA	Nordhaus
1975 Price: Busbar--$15/1,000 kWh[a] Delivered--$23/1,000 kWh		
Projected:		
1. Base Case		
1990	21.6	20.4
2000	21.6	19.7
2010	21.6	20.1
7. High GNP growth		
1990	21.8	20.5
2000	21.7	20.1
2010	22.0	20.7
8. Low GNP growth		
1990	21.6	20.2
2000	21.5	19.5
2010	21.5	19.6
9. Low GNP growth, nuclear moratorium, coal and shale oil limits, high-cost nonelectric AES		
1990	24.3	29.7
2000	21.7	22.1
2010	24.1	18.8

[a]Calculated by Alan Manne.

more likely to approach those projected here. Moreover, since all electricity prices reported here reflect both operating and annualized capital costs, they are, at least in concept, equivalent to long-run average costs, the pricing concept used currently by public utility regulatory commissions.

Finally, the primary resource price projections made by the models include economic rents, which are returns to resource owners in excess of the marginal costs of extracting the resources. In the language of mathematical optimization, these rents are the "shadow prices" on the individual resources. For example, although the cost of extracting domestic oil and natural gas is assumed to be $2 per million BTU's over the entire time horizon for each model, the reported prices for oil and

TABLE III.19B DESOM Projection of Prices of Electricity and Primary Energy Resources Under Different Policy and Realization Scenarios

Policies and Realizations: Base Case Modified as Stated	Oil ($/bbl)	Coal ($/$10^6$ BTU)	Uranium Minimum Acceptable Price ($/lb)	Electricity Delivered Consumer's Cost ($/1,000 kWh)
1975 Price	9.00	0.75	30.0	23.0
Projected:				
Base Case				
1990	14.39	1.17	--	35.7
2000	17.54	1.17	--	36.1
2010	23.22	1.17	--	36.9
Medium coal and shale limits				
1990	15.51	1.17	--	32.2
2000	21.58	1.17	--	38.8
2010	29.25	3.02	--	39.9
Low coal and shale limits				
1990	16.07	2.37	45.75	48.3
2000	22.36	1.88	50.29	47.9
2010	30.30	3.45	66.24	37.6
Medium coal and shale limits and nuclear moratorium				
1990	15.45	1.32	--	37.5
2000	21.35	1.97	--	44.6
2010	25.25	4.60	--	84.9
Low coal and shale limits and nuclear moratorium				
1990	13.77	2.01	45.0	44.5
2000	33.98	5.71	45.1	84.5
2010	26.15	4.69	--	74.0
Low coal and shale limits and low uranium availability				
1990	16.07	2.37	64.00	48.3
2000	22.36	1.88	114.80	47.4
2010	30.30	2.82	227.40	58.4

gas exceed this figure by an increasing margin over time, to reflect the increasing scarcity of the resource. The models indicate that these economic rents grow larger as the total remaining amount of the resource dwindles, as appropriate substitutes (e.g., synthetic oil derived from coal as a substitute for crude oil) become less available, and as the constraints on the use of the resource (e.g., limits on coal production) are more stringent.

In general, the price projections shown in Tables III.16 through III.19A and Figures III.9 and III.10 indicate approximately a doubling of oil prices and modest increases in the price of electricity through 2010 for most scenarios. However, as in the projections of total quantity of energy demanded, discussed in Section III.4, the effects of a nuclear moratorium *and* the imposition of coal and shale oil limits on all prices appear to be nonadditive. Thus, in this supply-restricted case, the increase in prices generally is greater than the sum of the increases that occur individually when each of these restrictions is imposed separately. Alternative GNP growth assumptions, on the other hand,

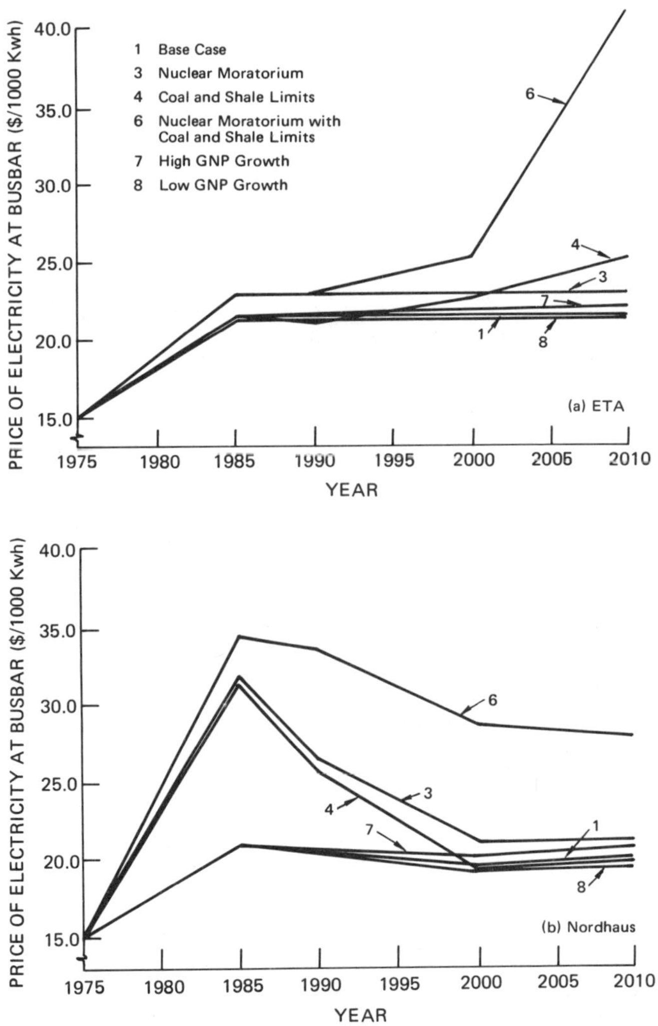

FIGURE III.9 Electricity price projections (in 1975 dollars).

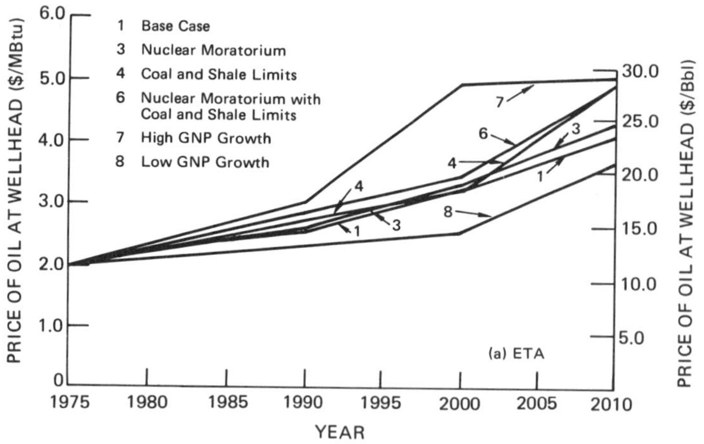

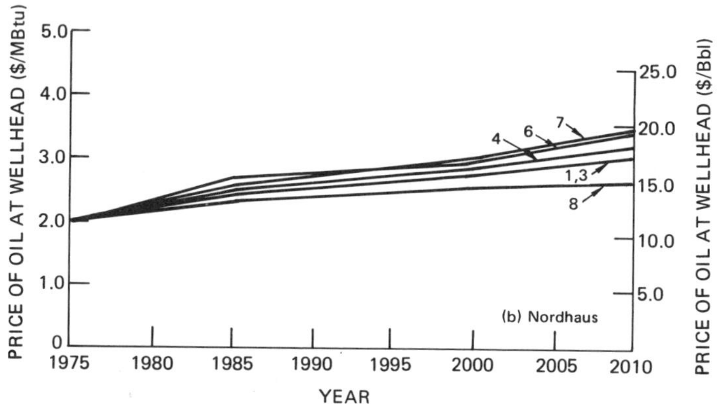

FIGURE III.10 Oil price projections (in 1975 dollars).

seem to have a much greater impact on the price of oil than on electricity prices in the absence of any energy supply system restrictions.

In comparing price projections between models, two results specific to particular models should be noted. First, in the early time periods the Nordhaus model reports higher electricity costs for some scenarios than the other models because of restrictions (specific to the Nordhaus model) on the short-run rate of capacity increase for individual technologies.

Second, for scenarios without coal and shale oil limits, ETA reports a slightly higher oil and gas price (about $0.40 per million BTU higher) in 2010 than in the years after that due to the MRG Base-Case restriction on the rate of introduction of new technologies. This "price hump" was not projected by the Nordhaus model, however.* More generally, the

*The price hump effect was more pronounced in earlier simulations where the rate of capacity increase in technologies was even more restricted than in the Base Case here (see Table II.3, Chapter II). On the other hand, in non-CONAES runs of ETA in which no constraints on the rate of capacity increase are present, the hump in price completely disappears.

oil prices projected by the Nordhaus model in 2010 are generally lower than those projected by DESOM and ETA due to three factors peculiar to the Nordhaus model: (1) lower income elasticities, (2) higher price elasticities, and (3) the allowance for greater direct burning of coal in Nordhaus than in ETA. In Section III.11, we attempt to assess the relative roles played by these three factors in causing differences in model results. Together, these factors tend to delay the depletion of low-cost oil and gas supplies (both domestic and foreign) in the Nordhaus model, which also postpones the large-scale introduction of synthetic fuels until after 2010.

ETA's electricity busbar (shadow) price projections roughly parallel those of the Nordhaus model (Tables III.18 and III.19A). Both show relatively flat projections between 1990 and 2010 at approximately $20/1,000 kWh for all scenarios except no. 6 (nuclear moratorium *and* coal and shale oil limits). Despite this flatness, all of these projections compare with busbar shadow prices in 1970 and 1975 of approximately $11/1,000 kWh and $15/1,000 kWh (in 1975 dollars), respectively,* which roughly approximate actual prices in those years.

As a result, the models agree on long-run projections of electricity prices that are somewhat higher than those currently prevailing. This conclusion is strengthened in those scenarios where coal and shale oil production are restricted. In those cases, competition for limited supplies of coal between producers of synthetic fuels and electricity-generating plants is projected by ETA to lead to a relatively larger increase in the price of coal than by Nordhaus, induced by the larger rent projected by ETA on coal production capacity (whose growth is constrained). This, in turn, causes ETA to project higher electricity prices than Nordhaus.

The DESOM projections of electricity costs are substantially higher for all scenarios than the projections made by ETA and Nordhaus. As Table III.18 indicates, this is because the ETA and Nordhaus projections refer to electricity at the busbar, while the DESOM projections refer to electricity delivered to the consumer.

On balance, the magnitude of the price increases projected in most of the scenarios is by no means alarming. Specifically, projected doubling of primary resource prices in some scenarios over the next 35 years implies annual real price increases of approximately 2 percent. Annual increases in electricity prices are projected to be even lower.

To test the sensitivity of these price projections to resource constraints of greater severity than specified either in the Base Case or in the medium coal and shale limits scenario, DESOM was used to project electricity and primary resource prices in scenarios incorporating the low coal and shale limit. These projections are reported in Table III.19B.

Several features of the projections are worth noting. As expected, coal prices rise sharply when the low coal and shale limit is imposed. Relative to the scenario of medium limits, coal prices reach their maximum in 1990, a year in which great pressure is placed on coal resources

*Computed by ETA.

as oil and gas resources will have been significantly depleted. By 2000, coal prices will have fallen from their 1990 maximum levels relative to the scenario of medium limits, since nuclear power assumes a much larger role in the generation of electricity, relieving, therefore, some of the pressure of demand for coal.

The coal prices projected for 2010 are absolutely and uniformly high for all scenarios involving coal and shale limits. They are highest when both a nuclear moratorium and low coal and shale limits are imposed. In the other scenarios coal prices are not as high because the presence of nuclear power removes some of the demand pressure from the limited supply of coal. Most surprisingly, coal prices in 2010 are lowest when both low coal and shale limits and low uranium resources are assumed, for then the role of the breeder is largest, reducing some of the upward pressure on the prices of other resources.

Oil prices, too, rise when the low coal and shale limit is imposed, but not as significantly as those of coal.

Electricity prices are somewhat less sensitive to the low coal and shale oil limit than primary resource prices because fuel costs form only a part of total electricity costs. The fuel cost displaying the greatest sensitivity is that of uranium. Although DESOM uranium price projections for the medium coal and shale oil limits scenarios are not available, the price projections reported in Table III.19B illustrate well how high uranium prices could rise in scenarios of significant resource scarcity. The presence of low coal and shale limits alone pushes the uranium price up to \$66.24/lb by 2010. If, in addition, uranium resource availability is low, then DESOM projects the much higher price of \$227.40/lb for that year.

The DESOM price projections, we recall, will tend to overstate the extent of price increases because the model does not permit demands for energy services to vary with energy prices. As a result, the DESOM projected prices tend to contain a larger economic rent component than prices projected by other models (see discussion of economic rent below). Nevertheless, the DESOM projections are useful in dramatizing the nature of the sensitivities and in placing a type of upper bound on the prices projected under the widely different scenarios considered here.

As indicated above, all price projections (for primary resources and electricity) include an allowance for economic rent, which arises when one or more supply constraints are binding. In some cases it is a laborious procedure to isolate what portion of a particular price is attributable to the rent component; in others, however, it is easily done. For example, because the MRG assumed coal extraction and delivery costs to be $\$0.75/10^6$ BTU (see Table II.1), all projected excesses over \$0.75 must represent positive rent. In the Base Case and a few other cases, coal prices (see Tables III.16 and III.17) are projected by ETA and Nordhaus to remain at \$0.75, implying no scarcity rents. But when either or both a nuclear moratorium and a limit on coal and shale are imposed, all models, to differing degrees, project coal prices above \$0.75 for some time period(s). The clearest case is when both restrictions are imposed simultaneously: The price projections for 2010 vary between \$1.47 (Nordhaus) and \$4.60 (DESOM). Prices at such levels (above \$1.50) include more rent (arising from the policy-imposed restrictions) than cost.

The oil price projections can be analyzed in much the same way. Since the MRG assumed that a fixed quantity of oil and gas resources (1,730 Quads) could be extracted at costs up to $\$2/10^6$ BTU, oil price projections above the dollar-per-barrel equivalent of that figure, or approximately $11.20/bbl, include a rent component. In contrast to coal, however, where any rents that arise are due to policy actions, the rents on oil rise specifically because of limited resource availability. Tables III.16 and III.17 indicate that rent plays an important role in the oil price projections for all cases for all models. The rent component in the projections ranges from 20 percent to nearly 60 percent of the price in some scenarios. In particular, as supply constraints are added or as GNP growth is increased, the projected prices, and therefore the rents, grow larger.

Furthermore, it is important not to overlook the fact that the size of economic rents on primary resources is strongly dependent upon the assumed cost of the nonelectric Alternative Energy System (AES). As indicated in Chapter II, the assumed cost of the AES acts as a price ceiling on other resources. The higher the assumed cost, therefore, the higher prices can be expected to rise in those scenarios where the price ceiling imposed by the AES actually prevents prices from rising still further. For example, ETA projects oil prices to rise to $29/bbl (equivalent to MRG Base-Case cost of the AES--$\$5/10^6$ BTU) by 2010 in all of the coal and shale limits scenarios (see Table III.16). Very nearly the same result is projected by DESOM. These projections suggest that even higher primary resource prices could be expected if the cost of the AES in 2000 is actually higher than the MRG Base-Case assumption.

Finally, the uranium price projections reported by ETA (Table III.16) indicate that the most rapid rate of uranium price increase is projected to occur in the scenario in which there are limits on coal and oil shale production *and* no FBR or ACR. This is a scenario that places the greatest strain on existing uranium resources through heavy reliance on the LWR, pushing the uranium price to a projected level of $61.70/lb by 2010. Uranium prices rise almost as much to reach $53.70/lb if the high GNP growth scenario is realized (Table III.17). Otherwise, in the Base Case or in the scenarios in which there are restrictions on nuclear power, the price of uranium is projected to remain relatively stable.

III.8 ECONOMIC BENEFITS AND COSTS OF ALTERNATIVE NUCLEAR REACTORS AND ECONOMIC COSTS OF ENVIRONMENTAL AND OTHER POLICY RESTRICTIONS

A central question addressed by CONAES and the MRG is what economic value there may be in developing nuclear technologies--whether advanced converters or breeders--that utilize uranium more efficiently than do the present Light Water Reactors. Equally important is the estimation of the economic cost of imposing restrictions on the use of nuclear fission and/or coal (even though possible environmental benefits or costs are not estimated in this report).

This section presents estimates that bear on both of these issues and are based on results of the six policy scenarios described in Chapter II. The R&D analysis of Chapter IV will supplement the nuclear reactor evaluations presented here with a more complete analysis, which explicitly incorporates probabilistic assessments of future values of some realization variables.

To provide the benefit and cost estimates here presented, the MRG turned to three of the long-run models, DESOM, ETA, and Nordhaus. Each of these models minimizes the discounted economic cost of meeting a set of energy demands over a long time horizon,* subject to a number of technological and behavioral constraints. In the DESOM model the path of energy demands is given *a priori*, which implies a price elasticity of demand equal to zero. In the ETA and Nordhaus models that path is itself obtained by maximizing the discounted sum of economic benefits to the consumer minus the costs incurred by the producer. The benefits are determined in a way already alluded to in the next-to-last paragraph of Section I.3. The estimation of the total benefits from energy consumption proceeds from the curve, called *demand function*, that represents the response of energy demand to energy price. For most commodities, including energy, this demand curve is downward sloping, indicating increases in demand as price falls. Empirically, a substantial number of investigators have confirmed this negative relationship between energy prices and quantities in a variety of energy-consuming sectors (see Table II.2). Roughly speaking, then, to estimate the increase in benefit due to an increase in energy use called forth by a small decrease in price, the increase in energy use is valued at that price (before its decrease, say†).

In all three models, as explained in Section I.4, the maximization of a suitable "objective function" can be seen as a simulation of the operation of a competitive market economy. It was also explained in Section I.4 that the technique of optimization facilitates the calculation of the maxima of the objective function attainable in the presence of any given set of technological options. Thus, in the DESOM model, differences in objective function values due to the presence or absence of particular technologies reflect differences in the total economic value of social resources (capital, labor, and primary energy) devoted to meeting the given set of end-use demands for energy services. In the other two models, the loss in real income of consumers when energy prices rise and hence energy use is correspondingly reduced is included as well.

We have used the term "real income" rather than GNP since the estimated benefit or detriment to the consumer from a change in energy use applies equally if total welfare is represented, inadequately, by GNP, or better by a measure of economic welfare (MEW) that takes leisure and household production into account. However, the interpretation of gain or loss in terms of real income, or net welfare gain, requires that as

*DESOM computes discounted costs through 2025, ETA through 2050, and Nordhaus through 2060.

†Since the estimations only involve comparisons between two different price levels (or energy use levels), the "constant of integration" cancels out in this procedure.

before we exclude from this term the evaluation of any environmental improvement or deterioration that may be associated with the change in energy use. As indicated in Section I, the state of the modeling art does not yet permit the incorporation of environmental and other noneconomic impacts in the objective functions of the models. These functions, therefore, reflect only the total *private* costs and benefits of producing and consuming energy. To compute the *social* cost of energy production one should then add judgmental estimates of the environmental impacts and other nonmarket costs that are associated with each projected supply and demand mix. Studies like those made by the Risk Panel may facilitate this difficult task.

It should also be noted that none of the three long-run models employed here (DESOM, ETA, and Nordhaus) account for regional differences in energy costs and demand. As a result, the objective function in each of the models must be looked on as a national aggregate.

The MRG took advantage of the optimization features of the three models just mentioned to estimate the economic costs (excluding R&D cost) of various exclusions or limitations of technologies hypothetically imposed in response to policy concerns. The results are recorded in Table III.20. All entries state the difference in maximal objective-function values of the Base Case over those of the constrained cases. Again, it should be noted that each of the models computes the objective-function values over a different time horizon: DESOM through 2025, ETA through 2050, and Nordhaus through 2060. This is likely to cause the estimates of DESOM to deviate somewhat from those of ETA and Nordhaus, but not so much as to deprive comparisons between model results of their meaning, because of strong discounting of the later outcomes.

That the price elasticities are substantially different between models is an important and informative source of discrepancies between estimates, however. In particular, one would expect DESOM, which does not permit the demands for energy services to be responsive to changes in price, to project much higher values of technology options or higher costs of technology restrictions than both ETA and Nordhaus, which assume that demands for energy are functions of energy prices. All of the results should be read with this difference in price elasticities in mind.

There are a variety of ways to read the table. One way is to analyze the net economic benefits (excess of benefits over costs other than R&D costs) of various nuclear options, each considered with or without concurrent limits on coal and shale oil production. Differences between these entries represent lump-sum equivalents of the discounted future gains in the economic value of goods and services that are estimated to become available if society gains access to, or removes a constraint from, the technology in question. The equivalent annual gains (in perpetuity) can be computed by multiplying the lump-sum figures by the discount rate used in these calculations, namely 6 percent.

In the Base Case and in most minor variations around it, the calculation finds that, if the use of coal and shale oil is not constrained, the economic benefits of both the ACR and the FBR will not, under the ETA and Nordhaus projections, cover projected Research, Development, and Commercialization (RD&D) costs, estimated to approximate $17 billion for the two programs combined (see Table IV.2). Even if only one advanced

TABLE III.20 Estimated Differences in Net Economic Benefits from Six Technology Mixes and Net Economic Costs of Five Alternative Policies to Reduce Environmental Impacts[a] (Unit: Billions of 1975 Dollars)

	DESOM (Costs Only)	ETA	Nordhaus
Shortfall below Base Case of benefits minus costs in six policy scenarios			
1. Base Case	(0)	0	0
2. Moratorium on all ACR's and the FBR[b]	(43)	8	2
3. Moratorium on all nuclear technologies	(105)	46	136
4. Coal and shale limits	(914)	159	64
5. Moratorium on all ACR's and the FBR and coal and shale limits	(1,012)	181	72
6. Nuclear moratorium and coal and shale limits	(2,325)	358	457

[a] In all policy scenarios total benefits and costs are the sums of year-by-year benefits and costs, discounted to 1975 at 6 percent per annum. DESOM computes only discounted costs, through 2025, ETA computes discounted benefits and costs through 2050, and Nordhaus through 2060. For each year, benefits estimate the value to the consumer of total amounts of energy consumed, on an incremental basis. For further explanations, see Section III.8.

[b] For ETA and SRI this policy includes a moratorium on the LWR with plutonium recycle; for the other models it does not.

NOTE: Costs of R&D are excluded.

reactor research program is pursued, the estimated RD&D costs (of between $7 and $10 billion) would still equal or outweigh the projected values of the "advanced nuclear" option.

Furthermore, the ETA specification of scenario 2 in Table III.20 (moratorium on all ACR's and the FBR) includes a moratorium on plutonium recycling; for Nordhaus it does not. As a result, ETA's projections of the cost of the moratorium, or alternatively, of the value of the advanced nuclear technology option, include the value of plutonium recycling--a process that allows current LWR's to economize on the use of uranium and is also a necessary accessory to advanced reactor implementation. ETA simulations in connection with the R&D programming work described in Chapter IV indicate the value of plutonium recycling in the Base Case to lie within the range of $4-$7 billion. This estimated value must be subtracted from the ETA Base-Case estimate in Table III.20 if the ETA estimated value of advanced reactors is to be compared with that of Nordhaus. The result of this exercise is an estimated Base-Case benefit of $1 billion to $4 billion of the advanced reactor package, a result in agreement with the Nordhaus projections. These two projections reinforce each other in the conclusion that, *if* the Base-Case values assumed by the MRG represent likely outcomes, the development of the advanced nuclear option by 2010 appears to have little economic rationale.

The projections made by DESOM, however, attribute a much higher value to successful development of an advanced reactor program--a discounted sum of $43 billion. This is the estimate of the lump-sum increase in the value of goods and services that society could reap from the advanced reactor package if energy services (automobile miles driven, space heating required, etc.) turn out to be insensitive to future changes in energy prices. As such, the estimate represents an "upper bound" on the benefits of advanced reactor technology in the Base Case, and also permits a test of their sensitivity with regard to the price elasticity of demand.

Estimation of the values of technological options by comparison with the Base Case is far from conclusive, however. A specific variation from the Base Case that was examined concerns the possibility that environmental restrictions may be enacted, which in effect will restrict coal and shale oil production. Since this would tend to increase the need for nuclear power, it is of interest how limits on coal and shale oil production would affect the benefit calculations for the package of advanced nuclear reactors, ACR's and/or the FBR.

A comparison of scenarios 4 and 5 reveals advanced reactor benefit estimates by Nordhaus and ETA of between $8 billion and $22 billion when ("medium") coal and shale oil limits are present. Again, the ETA estimate includes the benefits of plutonium recycling, which in a separate simulation were projected to be $17 billion in this scenario (see Table IV.2). Made comparable, therefore, the Nordhaus and ETA estimates of the value of further advanced nuclear R&D appear to lie, in the presence of coal and shale oil restrictions, somewhere between $5 billion (ETA) and $8 billion (Nordhaus). Considering again the R&D costs, the addition of coal and shale oil limits to the Base Case, therefore, does not appear to alter the ETA and Nordhaus findings on the economic benefits of advanced nuclear reactors.

Again, the DESOM projections show a substantial value of the advanced reactor program under the coal and shale oil limits. In this scenario, DESOM estimates the lump-sum economic benefit to reach $98 billion, which for reasons already discussed above should be treated as an upper bound on the value of advanced reactors in the presence of the coal and shale limits specified by the MRG.

It should be emphasized, however, that there are circumstances under which the economic benefits of the advanced nuclear option may be very substantial, also for models that allow demands to be responsive to price. Table III.21 presents alternative values of the FBR calculated by ETA under different realizations. It is extracted from Table IV.2 (to be discussed in Chapter IV) because it illustrates how the estimated benefits of one particular technology, in this case the FBR, are highly dependent on the values of certain realization variables. Specifically,

TABLE III.21 ETA Estimates of Net Economic Benefits of the FBR, Introduced in 2000, Under Different Realization Assumptions and in the Presence of Coal and Shale Limits[a]

	Uranium (U_3O_8) Availability		
	MRG Low (1.6 Million Tons at Prices up to \$45/lb; 2.25 Million Tons Total)	Intermediate (3.7 Million Tons at Prices up to \$45/lb; 3.7 Million Tons Total)	MRG Base (3.7 Million Tons at Prices up to \$45/lb; 12.57 Million Tons Total)
1. Low GNP growth			
Medium-cost AES	0	0	0
High-cost AES	0	0	0
2. Base-Case GNP growth			
Medium-cost AES	59	26	5
High-cost AES	123	76	NE[b]
3. High GNP growth			
Medium-cost AES	97	70	NE
High-cost AES	139	136	NE

[a]All simulations assume that the representative ACR is not available and that values of plutonium recycle are not included in the estimated values of the FBR. The AES refers to clean nonelectric AES's.

[b]Not estimated.

NOTE: Costs of R&D are excluded.

the table indicates that in the presence of coal and shale limits, estimated economic benefits of the FBR rise nonlinearly as estimated uranium availability falls. In some cases, notably high GNP growth and/or high-cost AES, the estimated benefits reach \$100 billion or more.

Of course, the FBR is one particular technology, and our sensitivity results span only a fraction of the total number of sensitivity simulations that could have been conducted. Nevertheless, the limited number of simulations reported in Table III.21 demonstrates the important finding that, for any technology, there will be some realization variables whose values are critical to the calculation of economic benefits. For the FBR (and the ACR), one of these variables is uranium availability; others are GNP growth, the presence of coal and shale limits, and the cost of a clean nonelectric AES. The list of critical variables will be similar, but not identical, for solar energy, electric or nonelectric.

As a result, the critical problem for R&D decisions is how to assess the probabilities of these various realizations. In the case of advanced reactor research, if the judgmental probabilities of the realizations leading to high-benefit cases are very small, then the advanced nuclear benefits conclusions from the Base-Case analysis will continue to hold. On the other hand, substantial probabilities of these high-benefit cases could tip the decision toward R&D on these nuclear options, even though their value is low in the Base Case. It is the purpose of the risk analysis in Chapter IV to clarify whether the existence of cases of high benefits could substantially change the initial findings reported here.

The analysis of the net economic benefits from nuclear fission need not stop with the advanced reactor question, however. The results for scenario 3 of Table III.20 report the projected value of the entire nuclear option including all types of reactors above the level of nuclear production already attained in 1974 (approximately 1.1 Quads). This value is projected by the three models to lie between $46 billion and $136 billion. If coal and shale oil limits are imposed, ETA and Nordhaus project the estimated value to rise to between $199 billion and $393 billion (the difference in results for scenarios 4 and 6). DESOM, however, projects, in the case of limited coal and shale, an upper bound value of the nuclear option of $1,411 billion, an amount that would be reaped if demands for energy services proved to be constant in the face of rising energy prices.

On a per capita basis (1976 population)* the estimates in Table III.20 imply that access to nuclear power as an energy option would provide an approximate $200 to $600 lump-sum increase, or the equivalent perpetuity, in terms of the value of goods and services available if no limits are imposed on coal and shale and if end-use demands are price sensitive; if coal and shale limits are imposed, the lump-sum per capita estimates rises to between approximately $900 and $1,800. For the hypothetical case in which end-use demands are not sensitive to prices, DESOM projects, in the limited coal and shale scenario, that nuclear power confers a lump-sum benefit of approximately $5,900 per person as an upper bound. This is not to say that the "average man" would reap such benefits. None of the models has attempted to project how the benefits would be distributed among individuals. To do so meaningfully would require introducing into the models such redistributive policy variables as may be available to the government to compensate for possible adverse distributive effects of otherwise beneficial energy policies.

A completely different way to read Table III.20 is to treat the estimates as gross economic costs (exclusive, that is, of savings in both environmental and R&D costs) of implementing certain restrictions. Thus, one can take the same dollar estimates that heretofore have been treated

*These per capita estimates are made by assuming 1976 population levels. Clearly, since population will grow in the future, it would ideally be necessary to calculate some type of weighted average over the 35-year period, 1975-2010, the weights being determined by some discount rate. To avoid the complexities of such a procedure, the 1976 population level was chosen simply to indicate the size of the costs in more manageable terms than discounted sums in "billions" of dollars.

as "values" or "benefits" of having a particular option and now treat them as "costs" of doing without that option. For example, just as one can say that the "value" of the nuclear option is projected to lie between $46 billion and $136 billion, one can also say that the economic "cost" of imposing a moratorium on new nuclear construction starts is projected to lie between the same figures.* In both of these cases, the calculation stops short of valuing any environmental cost of the technology, or any environmental benefit from its curtailment.

The cost aspect of these figures is brought out clearly by the impact estimates of coal and shale oil limits. Scenarios 4-6 of Table III.20 indicate that coal and shale limits alone are projected by ETA and Nordhaus to cost between $64 billion and $159 billion (a per capita lump-sum cost of approximately $300 to $750); with a nuclear moratorium, ETA and Nordhaus project the aggregate cost of the coal and shale limits to rise to between $308 billion and $393 billion (a per capita lump-sum cost of approximately $1,400 to $1,800). Finally, ETA and Nordhaus project a cost level of the nuclear and coal restrictions taken together (scenario 6) lying between $358 billion and $457 billion (a per capita lump-sum cost of approximately $1,700 to $2,100). Note that these cost estimates relate to limiting coal and shale to between three and four times current production levels. The DESOM scenarios given in Table III.19B suggest that restrictions of greater severity would cause the cost estimates to rise substantially.

As noted earlier, the DESOM cost estimates of coal and shale oil restrictions are much greater than the ETA and Nordhaus projections, because in DESOM end-use demands for energy services do not fall as prices rise. Consumers may substitute cheaper fuels for those that are more expensive (see Appendixes A and C), but the model does not, by assumption, permit the level of operation of end-use devices, be they cars, washing machines, or refrigerators, to vary with price changes. Consequently, when severe restrictions are placed on a single fuel, such as coal, the effects are substantial.

Table III.20 indicates that the cost of the medium coal and shale oil restriction could be as great as $914 billion (a lump-sum per capita cost of approximately $3,800) under the DESOM-imposed demand conditions. The cost of the combination of coal and shale oil restrictions and a nuclear moratorium could be over twice as high, $2,325 billion (a lump-sum per capita cost of over $10,000). It should be noted that the DESOM projections in Table III.20 are valuable primarily as sensitivity analyses demonstrating the crucial role of the price elasticity of demand

*The MRG did not estimate the economic cost of a total nuclear moratorium, which would include a shut-down of existing plants and a moratorium on all current construction. The cost of such a policy would be greater than the estimated cost of a moratorium on new construction starts, both because (1) other higher-cost Alternative Energy Sources would be needed to replace existing nuclear plants, and (2) local areas could suffer considerable costs of economic disruption if the transition from nuclear to Alternative Energy Sources is unforeseen and carried through in a brief period.

for energy in determining the economic values of technologies, or the economic costs of their curtailment.

We comment further on the ETA and Nordhaus valuations. In the coal and shale limits case, the ETA and Nordhaus models are projecting that, before the value of environmental protection is added back, society is projected to lose between $64 billion and $159 billion in the discounted value of goods and services that would otherwise be potentially available. One must weigh the benefits of reduced environmental impact and attainment of other policy goals in order to determine whether the imposition of such limits is socially desirable. Moreover, since the cost estimates in Table III.20 do not account for regional differences in energy costs and demands, one would have to make appropriate corrections to reflect geographical variations before arriving at any final conclusion.

On the whole, the estimates in Table III.20 reinforce conclusions drawn in other sections. For example, in Sections III.4-III.8, the models project some degree of quantity substitution between coal and nuclear energy if one or the other is restricted. Prices, therefore, are not projected to rise significantly in these scenarios. Both of these effects translate into the relatively moderate estimates, shown in scenarios 3 and 4 in Table III.20, of the costs of implementing one restricting policy at a time.

But when both coal and nuclear restrictions are imposed simultaneously (scenario 6), the cost estimates from individual restrictions are *not additive*. This is consistent with results in Sections III.4-III.7, which show substantial reductions in energy used and substantial increases in energy prices when both coal and nuclear power are restricted. Specifically, the cutbacks in quantitites and increases in prices are substantially greater than one would observe by adding the effects of the individual restrictions.

In closing this discussion, five important observations deserve attention.

First, one salient difference between the ETA and Nordhaus projections in Table III.20 not heretofore mentioned requires explanation. ETA projects a restriction on coal and shale to be more costly than one on nuclear power. For Nordhaus, comparison of cost projections has the opposite outcome. This difference is due, in large part, to the projection of substantial quantities of synthetic fuels in ETA and also to the restrictions ETA places on the coal-nuclear mix of electricity-generating plants. The synthetic fuels projection suggests a large cost of coal and shale oil limits and the restriction on the nuclear-coal plant mix reduces the cost of a nuclear moratorium.

Second, from a policy perspective, the estimates of the economic value of different nuclear technologies recorded in Table III.20 are the most important findings in this chapter. Despite the differences brought out here and elsewhere, it is noteworthy that the ETA and Nordhaus estimates of the value of nuclear power are so close. A similar observation can also be made about their price projections, although not as strongly.

Third, the estimates presented in this section should be read only in the context of the sensitivity analysis that accompanies them.

Table III.21, for example, illustrates the importance of GNP growth, uranium availability, imposition or nonimposition of coal and shale oil limits, and the cost of the AES's in assessing the estimated economic benefits of advanced reactor technologies. The DESOM results illustrate in stark form the importance of elasticity assumptions to all of the estimates, providing simultaneously a sensitivity test and an upper bound to the estimated costs for each of the scenarios. Further discussion of the impact of elasticity differences appears in Section III.11. In addition, the estimate of capital cost may be important for the evaluation of the capital-intensive FBR technology. There is uncertainty about the cost estimate with a chance for escalation of capital cost above the 1.25 of the LWR. The implications of an even higher capital cost for the FBR are straightforward. It would make the breeder less competitive, increase the economic costs of the constrained scenarios, and reduce the domain of outcomes over which the breeder can be introduced economically. Conversely a lower capital cost might make the FBR competitive even in the Base Case.

Fourth, it is possible that, on the basis of judgmental probabilities assigned to a wider range of cases than those reported here, the work reported in Chapter IV may qualify or modify the above conclusions.

Finally, the various estimates in Table III.20 of the economic costs of specific curtailments in energy use can also be seen as so many case studies of the energy-use-to-GNP feedback problem discussed in Section III.2. This thought will be followed up in the next section.

III.9 ESTIMATES OF THE FEEDBACK FROM ENERGY USE TO REAL INCOME

In this section we report on some model calculations that give quantitative background for the qualitative considerations of Section III.2 concerning the possible feedback from energy use to real income. We first give the results of some simulations made within the MRG study. Thereafter, we cite results of similar simulations made in other contexts.

There is still more to be learned from the simulation outcomes reported in Table III.20, using the interpretation suggested in the second heading of that table. The scenarios 2 through 6 all represent alternative policies that limit the supply of energy to the U.S. economy. Each of these does so by limiting the use of one or two available technologies, with various degrees of incisiveness, contemplated for environmental or other policy reasons.

For DESOM the entries for these scenarios represent, for each constraining policy, the increase (over the Base Case) the policy causes in the minimal discounted sum of year-by-year constant-dollar costs to achieve given paths of the various energy end-use services. For ETA and Nordhaus the entries represent the loss (below the Base Case) in the maximal discounted sum of year-by-year benefits-minus-costs of the energy-use paths generated by competitive energy markets, as simulated by that same maximization. These costs (DESOM) or losses of benefits-minus-costs (ETA and Nordhaus) cut into the real income available for nonenergy goods and services.

In Section III.2 we have argued that the long-run impact of gradual and foreseeable constraints on energy supply can be estimated on an assumption that the percentage of utilization of available capital and labor is equally high with and without the constraints. Following that line of reasoning, the entries in Table III.20 represent discounted sums of precisely the year-by-year impacts on total real income caused by the constraints--not counting the environmental and other gains that motivate the constraints. These impacts on real income are to be related to the curtailments of energy use caused by the same policies.

Figure III.11 below shows the outcomes by a scatter diagram of two

E on Horizontal Axis: Ratio of aggregate energy consumption in 2010 projected for each policy to that for the Base Case.

Y on Vertical Axis: Ratio of cumulative discounted GNP, 1975-2010, projected for each policy to that for the Base Case.

Symbols for Models are shown in the Legend. Projections for DESOM extend to 2025, for ETA to 2050, for Nordhaus to 2010. ETA and ETA (-0.5) have price elasticities of -0.25 and -0.5, respectively.

Symbols for Policies: B = Base Case for all models; for the lowest-energy ETA and ETA (-0.5) points the policy constrains energy use to 70 Quads throughout; for all other points see Figure III.12a.

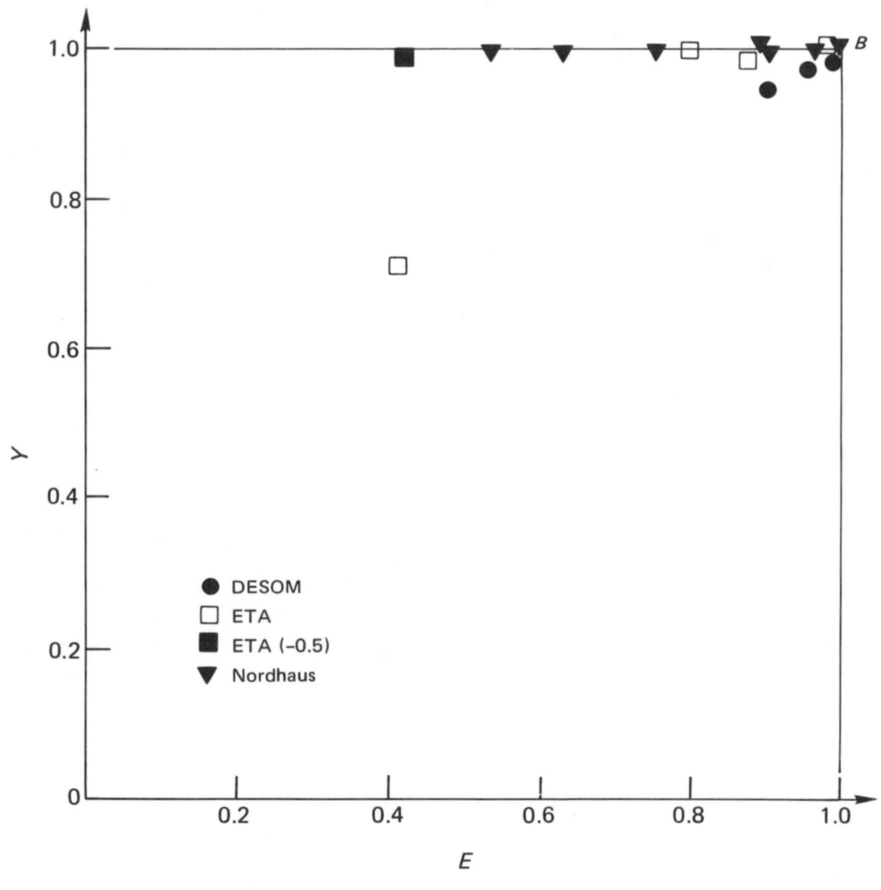

FIGURE III.11 Estimates of the long-run feedback from aggregate energy consumption on cumulative discounted real GNP.

ratios designed to make the results of the three models as comparable as their characteristics permit. The ratio set out along the horizontal axis has as its denominator the total energy use (in primary energy equivalents) projected for 2010 in the Base Case. In the numerator it has the same quantity projected for 2010 for the scenario in question.

The denominator of the ratio set out on the vertical axis is the discounted sum of the year-by-year GNP levels according to the MRG central values of the growth rate of GNP, given in Table II.1. The numerator is that same number minus the discounted sum of losses in real income, as defined above, associated with the scenario in question. Note that the numerator can be interpreted either as a loss of discounted real income or of discounted GNP, because the loss due to the constraining policy is the same for the two concepts. For the denominator, GNP was used because this was readily available. However, use of the Nordhaus-Tobin measure of economic welfare (MEW) for the denominator would have produced a ratio about halfway between unity and the ratio based on GNP, thus cutting the estimated feedback about in half. (This is due in large part to a valuation of leisure in MEW based on hourly labor income.)

Apart from the Base-Case point B = (1, 1), the points in Figure III.11 are labeled D_i, E_i, N_i for the three models, with subscripts referring to the scenarios. In addition, the figure shows two points labeled $E(-0.5)$ and $E(-0.25)$ that represent a sensitivity analysis using the ETA model* to test the effect of varying the price elasticity of aggregate demand for primary energy from -0.25 (the elasticity characteristic of the ETA model) to -0.50, under a zero-energy-growth policy. That policy holds the primary equivalent of energy use at a constant 70-Quad level through 1975-2050, while allowing optimization to determine the fuel mix used under that constraint at each point in time.

Two important observations stand out in inspecting Figure III.11. The first is the smallness of the feedback from cuts in energy use to real income. Apart from one DESOM point, D_6, and one ETA point, $E(-0.25)$, the feedback effect in terms of the ratio measure used is at most 2 percent, in spite of a more than 50 percent cut in energy use for the point $E(-0.5)$. In order to make these effects more visible, Figure III.12a reproduces the same points with the vertical scale stretched out by a factor of four. To provide a proper context for evaluating these losses: a 1 percent to 2 percent loss in real income in 2010 yet represents an undiscounted \$45 billion to \$90 billion of 1975 dollars. After discounting to 1975 at a 6 percent rate, it represents about 1.2 to 2.5 times the budget for the U.S. space program in 1975.

The second observation concerns the differences between the results for the three models, and for the two price elasticity values tested in the ETA model. Although the three models differ in various ways (see Appendix A), the most plausible explanation of the differences in results is in terms of price elasticities of demand for aggregate energy (measured in primary energy equivalents). Table III.22 shows the comparison of the elasticity numbers implicit in or built into the three

*Memorandum entitled "Some Implications of Zero Energy Growth," by Alan Manne, January 24, 1977.

Symbols for Policies: Numbered points in Figure III.12a refer to policies 3, 4, and 6 listed in Table III.20; the string of unnumbered Nordhaus points in both figures represents policies of successive curtailments of the growth rate of energy use, see Table III.13; the lowest-energy ETA point has run off the scale.

Y′ on Vertical Axis of III.12b: Ratio of undiscounted GNP in 2010 for each policy to that for the Base Case.

Other symbols as in Figure III.11.

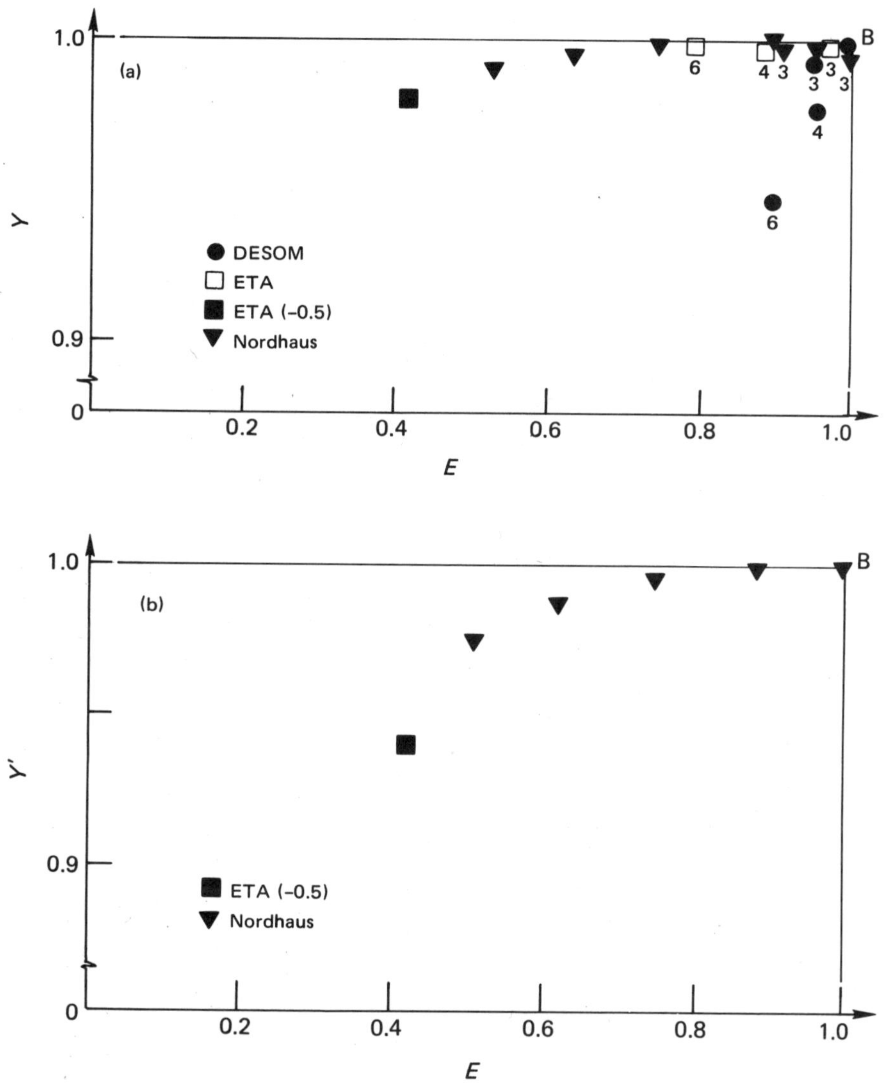

FIGURE III.12 Estimates of the long-run feedback from aggregate energy consumption on (III.12a) cumulative discounted GNP, 1975-2010, and (III.12b) undiscounted GNP for 2010.

TABLE III.22 Estimated Price and Income Elasticities of Demand for Aggregate Energy in Three Models

	DESOM	ETA	ETA(-0.5)	Nordhaus
Elasticity of demand for aggregate energy with respect to:				
Price	small[a]	-0.25	-0.50	-0.40[b]
Income	0.75	1	1	0.90[c]

[a]Since DESOM does not incorporate energy price responses by end-use consumers, its price elasticity reflects only the adjustment (small in absolute value) of the process mix between primary extraction and end use.
[b]Made comparable to ETA price elasticities by the procedure used in Appendix C, Table C.3.
[b,c]Average of elasticities measured at historical 1970-1972 prices and at 2010 prices of the Nordhaus model Base-Case projection; see Appendix C and Table C.3.

models. The DESOM points, calculated from a small price elasticity, show a feedback effect of up to 5 percent for policies that cut energy use out of the Base Case by not more than 10 percent. The ETA (-0.5) and Nordhaus points for the six scenarios, on the other hand, show feedback effects hardly more than 1 percent for energy cuts up to 20 percent. Finally, as the clincher, the two ETA points for a cut in energy use by 58 percent out of the Base Case give a feedback effect of only 2 percent if the price elasticity of demand is -0.50, and of as much as 29 percent if that elasticity is -0.25. These readings also suggest that the joint effect of price elasticity and of curtailment of energy use is nonlinear: The difference between the ETA (-0.25) and Nordhaus price elasticities hardly shows up in the reduced GNP figures for energy cuts up to 20 percent.

In contrast, the differences in time horizons between the three models are not believed to affect the comparisons much, for two reasons. First, since the discount rate (6 percent) exceeds the growth rate of GNP (3.2 percent through 2010, 1.8 percent thereafter) and that of energy use (less than 3 percent) by safe margins, the contribution to discounted sums from the parts of the horizon beyond 2010 is minor. Secondly, if the contribution beyond 2010 is chopped off, this should be done in both numerator and denominator, further reducing the effect on the ratio.

Table III.23 shows all the data plotted in Figures III.11 and III.12a. The table is subdivided into parts A and B, where in the scenarios of part A the cut in energy use is given, and the policy adjusts to it, while for those in part B, the policy is given and the energy cut follows from it. In both cases the question naturally arises how curtailments of energy use by the amounts implied in the various policy scenarios are to be brought about, should environmental and other noneconomic considerations dictate one or another of them. The MRG has *not* addressed this question as a possible policy problem. It has, however, utilized

TABLE III.23 Estimates of the Long-Run Feedback from Aggregate Energy Consumption on Cumulative Discounted Real GNP

Scenarios (Policies)	Implied Conservation Tax in 2010	Nordhaus A (35.0) / B (50.0)		ETA (49.4) Price Elasticity = -0.25		ETA (49.4) Price Elasticity = -0.50		DESOM (43.0)	
		$\bar{E}^a$	$\bar{Y}^a$	$\bar{E}^a$	$\bar{Y}^a$	$\bar{E}^a$	$\bar{Y}^a$	$\bar{E}^a$	$\bar{Y}^a$
A. Upper limits on growth rate of energy consumption, percent per annum	$(75)/$10^6$ BTU								
None	0	1.000	1.000	1.000	1.000	1.000	1.000	--	--
1.5	0.34	0.893	0.999	--	--	--	--	--	--
1.0	1.02	0.751	0.997	--	--	--	--	--	--
0.5	1.98	0.631	0.995	--	--	--	--	--	--
0.0	3.19	0.531	0.991	--	--	--	--	--	--
Upper bound of 70 Quads	[b]	--	--	0.419	0.711	0.419	0.982	--	--
B. MRG scenario (see Table III.20)									
1. Base Case	--	1.000	1.000	1.000	1.000	--	--	1.000	1.000
3. Nuclear moratorium	--	0.968	0.997	0.980	0.999	--	--	0.984	0.998
4. Coal and shale limits	--	1.000	0.999	0.894	0.997	--	--	0.952	0.979
6. Nuclear moratorium and coal and shale limits	--	0.901	0.991	0.795	0.993	--	--	0.900	0.946

[a] $\bar{E}$ = Ratio of aggregate energy consumption in 2010 in the indicated scenario to that of the Base Case.
$\bar{Y}$ = Ratio of cumulative discounted GNP in the indicated scenario to that of the Base Case.

[b] Price elasticity, -0.50 only; tax for electricity, 126 mills/kWh; oil and gas, 8.9 $/10^6$ BTU.

NOTE: Number in parentheses after name of model indicates total discounted GNP, in trillions of 1975 dollars, calculated over the horizon of the model.

the idea of a hypothetical "conservation tax" on primary energy for the scenarios in part A. This is a computationally convenient device for fitting the assumed energy cuts into models that are constructed on the basis of consumers' and producers' quantity responses to prices. This "tax" could be defined as the excess of the price, needed to reduce demand for energy to the curtailed level, over the price sufficient to call forth the supply of that curtailed amount. We use the term in that sense here, although in actual policy any non-price-induced reduction in energy use that can be mandated or otherwise brought about will correspondingly reduce the tax actually needed to balance supply and demand.

Part A of Table III.23 also contains the results of an experiment with the Nordhaus model,* in which successively more severe upper limits are imposed on the growth rate of energy consumption from 1975 on, starting with no limit, then descending from a limit of 1.5 percent per annum by 0.5 percent steps, ending in a zero-growth constraint. The results appear in Figures III.11 and III.12 as a string of triangular points (those without policy labels in Figure III.12a), called the "Nordhaus frontier." For both the ETA and Nordhaus projections, Table III.23, part A, also shows the conservation tax that would be needed to "close the market" in the absence of non-price-induced conservation. Finally, Figure III.12b shows a Nordhaus frontier and one ETA (-0.5) zero-energy-growth point in terms of the ratio of undiscounted GNP in 2010 in the energy-growth-constrained case to that in the Base Case. The data are given in Table III.24.

There is an interesting reason why all the Nordhaus points that correspond to scenarios in part B of the table fall below the Nordhaus frontier in Figures III.11 and III.12a, a fact more clearly visible in Figure III.12a, where the vertical scale has been stretched out. A similar result would have been obtained if similar frontiers had been calculated for the DESOM and ETA models. Because all of these are optimization models, the frontier estimates the minimum feedback from energy use to the sum of discounted GNP that can be achieved by any mixture of technologies satisfying the growth constraint in question. As said above, the curtailment of energy use here under discussion is not an end in itself, but a stand-in for more specific cuts that may be applied to specific technologies deemed possibly harmful or otherwise undesirable. For this reason, the points D_i, E_i, and N_i are more representative of feedbacks from specific constraints on energy supply than is the frontier.

The MRG results about the extent of the feedback confirm earlier results by other investigators and add further insight into the effect of the price elasticity of demand for energy. The first feedback study based on an econometric model, out to the year 2000, was made by Hudson and Jorgenson† and presented elsewhere by the same authors‡ in greater

*"What is the 'Tradeoff' Between Energy Consumption and Real Income?," W. Nordhaus, November 1, 1976.

†Appendix F, "Economic Analysis of Alternative Energy Growth Patterns, 1975-2000," in *A Time to Choose*, Ballinger Publ. Co., 1974.

‡"U.S. Energy Policy and Economic Growth, 1975-2000," *The Bell Journal of Economics and Management Science*, vol. 2, Autumn 1974, especially sec. 6.

TABLE III.24 ETA and Nordhaus Estimates of the Feedback from Energy Consumption on Real GNP in 2010

Energy Consumption Levels $(\bar{E})^b$	Nordhaus $(\bar{Y})^b$	ETA Primary Energy Price Elasticity = -0.5)[a] $(\bar{Y})^b$
1.000 (Base Case)	1.000	1.000
0.893	1.000	--
0.751	0.996	--
0.631	0.989	--
0.531	0.977	--
0.419	--	0.942

[a] Due to the approximation technique used in ETA, corresponding figures for the case where ETA's primary energy price elasticity is set at -0.25 cannot be provided.

[b] $\bar{E}$ = Ratio of aggregate energy consumption in 2010 to that of the Base Case.

$\bar{Y}$ = Ratio of real income in 2010 for each level of energy consumption to that of the Base Case ($4,500 billion).

detail for the years 1980 and 1985, writing in 1974. Using again the conservation tax device (they refer to it as a "BTU tax"), they project in the latter study that a tax of $0.50 per million BTU in 1980, compressing total energy input by 7.8 percent, decreases GNP by 0.42 percent, all relative to a no-tax Base Case.

A few differences between the assumptions underlying these estimates and those made in the applications of ETA and Nordhaus above should be noted. The Hudson-Jorgenson (H-J) model is an equilibrium model covering the entire economy (with four nonenergy sectors and five energy sectors). This aspect of the H-J model allows a fuller tracing of the impacts of a tax-induced curtailment of energy use in various energy-using industries, giving quantitative background for the general discussion of substitution of capital and labor for energy in Section III.2 above.

For comparing the conservation tax rates of the H-J, ETA, and Nordhaus models, it should be kept in mind that the H-J tax is levied on the BTU content of energy as it leaves the energy sector for use by other sectors. In the ETA and Nordhaus analysis, the tax is implicitly levied on the use of primary energy or energy equivalent. One can view this as a levy on BTU content at the point of entry into the energy-producing and energy-converting sector.

It should be emphasized again that a tax on energy as such, whether primary or ready for use, is not a suitable device for balancing economic costs of curtailment of energy use against environmental or other policy gains. Different fuels are environmentally harmful in different ways and degrees. Therefore, if a conservation tax is actually to be used, it should be applied in different degrees attuned to the specific detriment or risk to be minimized. In the present discussion we have considered only two extremes, policies that cut down on one or two technologies, or policies that operate through a blanket tax on all forms of energy. A realistic tax proposal to achieve the ends in question would have to be fine-tuned to its purposes.

Summing up, we estimate that the negative feedback from a curtailment of energy use below the Base-Case path is moderate, in fact small, unless *both* the curtailment is large--say by 50 percent--*and* the price elasticity of demand is small in absolute terms--say substantially below one-half.

Some further remarks should be addressed to the comparison of this finding with other work in the CONAES study, or in progress elsewhere.

First, the MRG assumption of a GNP growth rate averaging 3.2 percent over the period 1975-2010 is based on projections of growth in population, changes in labor-force participation, increases in productivity, and private decisions and public policies that influence the composition of each year's GNP between consumption and investment. Should it be found that reduced growth in energy use will in some way influence this allocation so as to decrease the amount of savings that go into capital formation, then on that ground alone the GNP in 2010, and the portion of the GNP path approaching that year, would have to be somewhat lower. The MRG estimates are based on an assumption that such an effect either can be disregarded, or can be counteracted by government policy likely to be implemented.

Second, the reasoning of this section is based entirely on the estimated response of energy demand by consumers and producers to energy prices. How is this to be squared with Schurr's observations* on the historical record of the period 1920-1945? In this period, technical changes due to the substitution of more flexible and refined energy forms (gasoline, electricity) for forms closer to primary energy had a positive effect on the ratio of GNP to primary energy use clearly dominating the negative effect attributable to the (then) declining prices.

The MRG projections are based on an assumption of a rate of technical progress in energy extraction, conversion, and use equal to that in the economy in general--although historically the progress in the energy sector has been faster. It is quite possible that energy-related technology will again, through its own as yet only conjectured opportunities for innovation, open new pathways that go beyond the technical options spelled out in the MRG projections, and beyond the past response to price change. One may think, for instance, of the fuel cell, the heat pump, and the use of hydrogen as a fuel or a base for fuels, as possible candidates for extending the contributions of liquid fuels and electricity. Should such spurts recur in the coming 35 years, then the effect would be that the moderate or small negative feedback on GNP from

*Sam H. Schurr, "Historical Relationships Between Energy Use and GNP in the U.S.," revised draft, February 15, 1977, typescript.

a given curtailment of energy use would be diminished, or even again dominated by, the increase in GNP due to innovations in the form of energy used, in spite of the curtailment of aggregate use of primary energy.

Third, the foregoing estimates are all based on particular models embodying empirical information. Valuable conceptual and analytical discussions that support and clarify the findings of this section can be found in a paper by W. Hogan and A. S. Manne,* and in the paper by Nordhaus already cited.

III.10 CAPITAL SHORTAGE POSSIBILITIES

An issue of potential concern is whether sufficient capital will be available to the energy-supply and processing industries in future years. Should this sector demand an unusually large share of future total investment, interest rates for all borrowers would be likely to rise. They would rise even more for energy-sector borrowers attempting to implement new processes perceived as carrying unusual risks. To the extent this occurs, investment in the energy-supply sector would be retarded, thereby producing higher prices and hence lower demands than those projected in this report.

The 1976 annual report of the Federal Energy Administration† summarizes the issues related to the likelihood of a capital shortage in the energy-supply industries and presents the results of several recent studies that give conflicting answers to the capital shortage question. From these estimates, one can infer that investment in the energy-supply sector and the electricity sector represented 2.64 percent and 1.25 percent, respectively, of the GNP in 1974.

The Approach

Cumulative investments in the energy-supply sector as a whole and in the electricity sector in particular are projected by the ETA model for each scenario. Insight into the possibility of a capital shortage in these sectors can be gained by comparing the ratios of these cumulative investment projections to the cumulative GNP projections that are assumed for each scenario with the ratios that prevailed in 1974. Special attention should be given to the electricity sector because the trend towards electrification in the future, projected in Section III.6, coupled with the capital-intensive nature of the industry, would tend to make a capital shortage surface there before it would in the other energy-supply industries. Additionally, since the ETA model projects a

*Energy-Economy Interactions: The Fable of the Elephant and the Rabbit? Working Paper EMF 1.3, Energy Modeling Forum, Stanford University, January 12, 1977.

†1976 National Energy Outlook, p. 293.

greater rate of growth in the electricity and synfuels industries (which are relatively capital-intensive) than the Nordhaus model, any potential capital shortage is more likely to show up in ETA scenario projections.

It should be noted that these projections do not include the capital acquisitions of energy users who may find that their most economic alternative to high energy prices is to purchase energy-utilizing equipment that is more energy-conserving but also more expensive. The work of the Demand/Conservation Panel represents one of the first attempts to project such investments, and should be given attention equal to the supply-side investment projections reported here.

In addition, the projections also do not include possible capital costs associated with the nonelectric Alternative Energy System (AES), principally because no particular technology or resource has yet been designated as the AES. For most scenarios, the lack of capital cost data for the AES makes little difference to the investment projections. In the scenario where coal and nuclear power are restricted simultaneously, however, the AES can play a significant role by 2010 in meeting energy demands. The investment projections for this scenario, therefore, may understate the investments that may actually be required if coal and nuclear restrictions are actually adopted.

The Results

Tables III.25A and III.25B show the projections for 5- and 10-year periods of (undiscounted) supply-side investments in the electric and nonelectric energy sectors as a percentage of GNP for the various policy and realization scenarios. The projections show a drop below the 1974 percentages in the near term with a return to those levels by about 2010 for most scenarios. These projections are the result of two counteracting trends: (1) the energy price rise coupled with the price responsiveness of energy demands, which tend to result in lower energy growth in relation to GNP growth through the entire time horizon, and (2) the requirement for relatively capital-intensive synfuel and nuclear power plants after the turn of the century.

High GNP growth tends to increase capital requirements (as a percentage of GNP) in the energy sector, while low GNP growth tends to decrease them (see Table III.25B). At the same time, GNP growth is a realization variable and though high GNP growth could strain the ability of the energy sector to expand, it still should have significant positive effect on our future standard of living. Yet, even with high GNP growth, a significant capital shortage in the energy-supply sector appears unlikely.

III.11 EXPLAINING INTERMODEL DIFFERENCES

Since different policy conclusions can sometimes be drawn from the scenario results projected by the alternative models, it is important to understand what factors may be responsible for these differences. The explanatory task is a difficult one, however, because of the diversity in model structures among the models used by the MRG.

TABLE III.25A ETA Projections of Undiscounted Energy- and Electric-Sector Investment as a Percentage of GNP Under Different Policy Scenarios (Cumulative GNP and Investment Totals for Each Period Indicated in 1975 Dollars)

Policies: Base Case, Modified as Stated (Scenario Number Indicated at Left)	GNP	Energy-Sector Investment		Electric-Sector Investment	
	($ × 10^{12})	($ × 10^{12})	(% GNP)	($ × 10^{12})	(% GNP)
1974	--	--	*2.6*	--	*1.25*
1. Base Case					
1976-1985	19.8	0.46	2.3	0.17	0.85
1986-1990	12.9	0.26	2.0	0.12	0.95
1991-2000	32.4	0.62	1.0	0.37	1.13
2001-2010	42.4	1.06	2.5	0.50	1.17
2. Moratorium on all ACR's and the FBR					
1976-1985	19.8	0.46	2.3	0.17	0.85
1986-1990	12.9	0.26	2.0	0.12	0.95
1991-2000	32.4	0.62	1.9	0.36	1.11
2001-2010	42.4	1.05	2.5	0.49	1.15
3. Nuclear moratorium					
1976-1985	19.8	0.44	2.2	0.15	0.75
1986-1990	12.9	0.25	1.9	0.11	0.85
1991-2000	32.4	0.59	1.8	0.34	1.05
2001-2010	42.4	1.02	2.4	0.45	1.06
4. Coal and shale oil limits					
1976-1985	19.8	0.46	2.3	0.18	0.91
1986-1990	12.9	0.27	2.1	0.13	1.01
1991-2000	32.4	0.59	1.8	0.32	0.99
2001-2010	42.4	1.18	2.8	0.53	1.25
5. Moratorium on all ACR's and the FBR, coal and shale oil limits					
1976-1985	19.8	0.46	2.3	0.18	0.91
1986-1990	12.9	0.26	2.0	0.13	1.01
1991-2000	32.4	0.59	1.8	0.32	0.99
2001-2010	42.4	1.18	2.8	0.52	1.23
6. Nuclear moratorium, coal and shale oil limits					
1976-1985	19.8	0.44	2.2	0.16	0.81
1986-1990	12.9	0.25	1.9	0.12	0.93
1991-2000	32.4	0.56	1.7	0.29	0.90
2001-2010	42.4	1.11	2.6	0.21	0.50

TABLE III.25B ETA Projections of Undiscounted Energy- and Electric-Sector Investment as a Percentage of GNP Under Different Realization Scenarios (Cumulative GNP and Investment Totals for Each Period Indicated in 1975 Dollars)

Realizations: Base Case Modified as Stated (Scenario Number Indicated at Left)	GNP	Energy-Sector Investment		Electric-Sector Investment	
	($ × 10^{12})	($ × 10^{12})	(% GNP)	($ × 10^{12})	(% GNP)
1974	--	--	*2.6*	--	*1.25*
1. Base Case					
1976-1985	19.8	0.46	2.3	0.17	0.85
1986-1990	12.9	0.26	2.0	0.12	0.95
1991-2000	32.4	0.62	1.9	0.37	1.13
2001-2010	42.4	1.06	2.5	0.50	1.17
7. High GNP growth					
1976-1985	21.4	0.59	2.8	0.26	1.22
1986-1990	15.3	0.38	2.5	0.21	1.37
1991-2000	41.9	1.10	2.6	0.75	1.79
2001-2010	62.0	1.88	3.0	0.80	1.02
8. Low GNP growth					
1976-1985	18.1	0.35	1.9	0.10	0.55
1986-1990	11.2	0.21	1.9	0.09	0.69
1991-2000	25.1	0.37	1.5	0.20	0.80
2001-2010	27.2	0.40	1.5	0.18	0.66
9. Low GNP growth, nuclear moratorium, coal and shale oil limits, high-cost nonelectric AES					
1976-1985	18.1	0.34	1.9	0.10	0.55
1986-1990	11.2	0.20	1.8	0.08	0.71
1991-2000	25.1	0.38	1.5	0.21	0.84
2001-2010	27.2	0.37	1.4	0.11	0.40

Given the extensive use made of three of the long-run models, DESOM, ETA, and Nordhaus, the MRG focused its attention on the intermodel differences among these particular models. Furthermore, within this subgroup of models, only DESOM assumes zero price elasticities for end-use demands for energy services. As a result, the DESOM results have been viewed as an approximate upper bound on the price and quantity projections for each scenario. They also serve as an indirect sensitivity test of the effect of nonzero price elasticities of demand. Of course, there are cases where this view is not quite correct because other differences between models may sometimes overwhelm those generated by elasticity differences. Nevertheless, to a large degree, the DESOM results are distinguishable from those of ETA and Nordhaus. For this reason, we concentrate here only on the ETA and Nordhaus models and attempt to attribute differences in the projections of total primary energy consumption in 2010 made by ETA and Nordhaus to differences in:

(1) assumed income elasticities of energy demand,
(2) assumed aggregate price elasticities of energy demand, and
(3) other facets of the representations of the energy system embodied in the models, including primarily differences in assumed cross-price elasticities of demands between specific fuels.

Appendix C documents and attempts to compare the price and income elasticities embodied in all of the models. Since elasticity comparisons, however, turn out to be quite difficult to make, the results reported here were derived using a method explained in Appendix D, which does not require the direct comparison of the elasticities assumed in the various models.

Table III.26 reports differences in total primary energy consumption projections made by ETA and Nordhaus for 2010. These differences are then attributed to differences in the above three categories of model assumptions. For example, approximately 26 percent of the 49.1-Quad Base Case difference in primary energy consumption projections for 2010 can be attributed to differing assumed income elasticities, 12.8 percent to differing assumed aggregate price elasticities for aggregate energy demand, and the remaining 60.7 percent primarily to differing assumed cross-price elasticities of demands between specific fuels. Differences in the alternative assumptions about the cross-elasticities of demand between the various energy fuels do not affect the projected amounts of fuels actually received by energy consumers; they do, however, explain most of the sizable differences in energy conversion losses (in electricity generation and synthetic fuel production) projected by the two models.

The results reported here show that differences in all three categories of model assumptions are significant in explaining the differences in primary energy consumption projections made by the models under every scenario. In particular, the absolute difference explained by the differing income elasticity assumptions is about the same for all the policy scenarios (i.e., those with the Base-Case GNP growth assumption). Consequently, when coal and shale limits are imposed, reducing in ETA the production of synthetic fuels, the percentage difference attributable to differing income elasticity assumptions doubles from close to 25 percent in the Base Case to approximately 50 percent.

The differences in cross-elasticity assumptions are primarily due to the allowance of greater direct coal burning by industry in the Nordhaus model (where industrial coal use is essentially unlimited) than in the ETA model (where direct use of coal is limited to 9 Quads in 2010). This

TABLE III.26 Estimated Sources of Differences in the Total Primary Energy-Consumption Projections for 2010 Made by the ETA and Nordhaus Models Under Different Scenarios (Unit: Quads of Primary Energy Equivalents)

	Primary Energy Consumption			Sources of Difference		
Scenario	ETA	Nordhaus	Difference	Income Elasticities	Aggregate Price Elasticities	Other[a]
1. Base Case	178.9	129.8	49.1	13.0 (26.5%)	6.3 (12.8%)	29.8 (60.7%)
2. Moratorium on all ACR's and the FBR	178.4	129.6	48.8	12.0 (26.6%)	6.3 (12.9%)	29.5 (60.5%)
3. Nuclear moratorium	175.4	125.5	49.9	12.5 (25.1%)	10.4 (20.8%)	27.0 (54.1%)
4. Coal and shale oil limits	160.0	133.1	28.9	13.3 (49.0%)	4.3 (16.0%)	9.3 (35.0%)
5. Moratorium on all ACR's and the FBR, coal and shale oil limits	159.3	131.7	27.6	13.3 (48.0%)	5.2 (19.0%)	9.1 (33.0%)
6. Nuclear moratorium, coal and shale oil limits	142.3	116.8	25.5	11.6 (45.0%)	9.2 (36.0%)	4.7 (19.0%)
7. High GNP growth	256.4	199.9	56.5	30.2 (53.5%)	8.1 (14.3%)	18.2 (32.2%)
8. Low GNP growth	98.8	82.3	16.5	3.6 (22.0%)	6.1 (37.0%)	6.8 (41.0%)
9. Low GNP growth, no nuclear moratorium, coal and shale oil limits, high-cost nonelectric AES	99.5	83.0	16.5	3.5 (21.0%)	3.4 (21.0%)	9.6 (58.0%)

[a] Includes primarily differences attributable to different cross-elasticities of demands for alternative fuels and differences in methods of energy product aggregation.

leads to a lower synthetic fuels requirement in the Base Case in the Nordhaus projections, as noted above, and implies a lower cross-elasticity of the demand for electricity with respect to nonelectric energy fuels. The latter, in particular, explains the remaining cross-elasticity effect in the scenarios with coal and shale oil limits.

Finally, the impact of the alternative income elasticity assumptions is, not unexpectedly, positively correlated with the GNP growth assumption, while the effect of different aggregate energy price elasticities, not surprisingly, is largest when energy prices are the highest.

Table III.26, therefore, highlights the importance of elasticities in explaining, to a significant degree, the differences between the ETA and Nordhaus projections of aggregate energy production. As a further test of the importance of the elasticity assumptions, several additional policy scenarios were simulated by ETA under two different sets of own price and income elasticity assumptions. The results of this experiment are presented in Table III.27.

Because the simulations reported in Table III.27 were made at a different date and for a different purpose than those reported earlier,* the Base Case and other policy scenarios reported in the table contain a number of important differences in assumptions from those made in the MRG Base Case. Among them, these are the most important:

1. Uranium resources of 3.7 million tons are available at a minimum acceptable price of $45/lb of U_3O_8. No resources are available at higher cost (in contrast to the MRG Base-Case uranium availability of a total of 12.57 million tons at a price of up to $225/lb U_3O_8).

2. No advanced converter is included as a technology option (the opposite is true for the MRG Base Case).

3. The average annual growth in GNP through 2010 is assumed to be 3.06 percent (to be compared with the MRG Base-Case assumption of 3.19 percent).

4. No non-price-induced conservation is assumed (to be compared with the MRG Base-Case assumption that a 10 percent non-price conservation effect is achieved by 2010).

Other than these deviations, the ETA simulations and definitions of policy scenarios are the same as those used throughout this report. The last scenario reported in Table III.27, however, includes a 15 percent non-price-induced conservation effect, which exceeds the 10 percent figure assumed in the MRG Base Case.

The top of the table reports the two sets of elasticities used in the comparison. Set A includes a lower price elasticity (-0.25) and a higher income elasticity (1.00) than set B (-0.50 and 0.80 respectively). While it would have been most useful to compare results when only one elasticity is varied at a time, limited time and resources have restricted our attention to the case where the two elasticities are varied together.

*These simulations were made at the request of the Inter Panel Scenario Group, an arm of the Synthesis Panel of the CONAES study. Several of the MRG Base-Case assumptions were changed for this purpose.

TABLE III.27 ETA Results for Selected Scenarios and Output Variables for 2010 Under Two Alternative Sets of Price and Income Elasticity Assumptions

Elasticity Assumptions	Set A	Set B
Long-Run Price Elasticity of Demand for Primary Energy	-0.25	-0.50
Long-Run Income Elasticity of Demand for Primary Energy	1.00	0.80

	Energy Prices				Energy Consumption					
	Oil and Gas ($\$/10^6$ BTU)		Electricity at Busbar (Mills/kWh)		Coal Consumption (Quads)		Total Primary Energy Consumption (Quads)		Electricity Generation (10^{12} kWh)	
Elasticity Assumption Set	A	B	A	B	A	B	A	B	A	B
Scenarios:										
1. Base Case[a]	4.5	3.5	21.7	21.5	101.7	41.5	186.4	120.0	7.6	4.4
2. Coal and shale limits	5.0	3.4	25.0	22.4	45.0	36.5	167.3	119.7	7.3	4.3
3. Coal and shale limits, nuclear moratorium	5.0	3.7	66.0	35.8	45.0	45.0	150.7	108.9	3.8	3.1
4. Coal and shale limits, nuclear moratorium, high-cost AES, 15 percent non-price-induced conservation	8.0	3.3	71.0	27.7	45.0	45.0	122.3	101.1	3.8	3.1

[a] See text for definition.

Nevertheless, the results are highly instructive. In particular, in scenarios 1-3, where supply constraints are only moderately severe, set B elasticities produce substantially lower estimates of all quantities reported and of the prices of oil and gas. Electricity prices are relatively insensitive to differences in elasticities in the Base Case and in the presence of coal and shale limits, but can be heavily influenced by elasticities when both coal and shale limits and a nuclear moratorium are imposed.

To supplement these results, Table III.28 reports, for the same set of assumptions used in Table III.27, projections of the differences in the value of the ETA objective function (measured relative to the Base Case) generated by two policy scenarios under the two sets of elasticity assumptions, sets A and B. The figures in the table can be interpreted in exactly the same two ways as Table III.20: (1) as measures of the lump-sum discounted value of having access to a particular technology and (2) as the strictly economic cost (exclusive of the cost of environmental impact) of imposing the indicated restrictions.

The table indicates substantial differences in the gross cost (or benefit) estimates between the two sets of elasticity assumptions. In part, this is because of the assumption in these simulations of uranium resource availability well below that of the MRG Base Case. Nevertheless, the large part of the difference is evidently attributable to the highly nonlinear impact of the elasticity differences.

The results in Tables III.27 and III.28 have an important bearing on the differences in results reported by ETA and Nordhaus. As Table III.22 and Appendix C indicate, the price elasticities in the Nordhaus

TABLE III.28 ETA Estimates of the Economic Cost of Two Technology Restrictions Under Two Sets of Elasticity Assumptions

Elasticity Assumptions:	*Set A*	*Set B*
Long-Run Price Elasticity of Demand for Primary Energy	-0.25	-0.50
Long-Run Income Elasticity of Demand for Primary Energy	1.00	0.80

	Difference in Cumulative Discounted Costs (in Billions of 1975 Dollars, and at a Discount Rate of 6 percent)	
Elasticity Set	A	B
Policies:		
Coal and shale limits	196	36
Coal and shale limits and nuclear moratorium	533	117

model are between 3/2 and 2 times those built into ETA, and the income elasticities appear to average somewhere between 0.8 and 0.9 times the unitary ETA income elasticity. The ETA exercise just presented, therefore, is a rough empirical test of what results ETA would have generated had it used the Nordhaus price and income elasticities. The striking aspect of the results is that the B set of elasticities (which approximate those of Nordhaus) produced projections of quantities, prices, and, to some degree, economic costs (or benefits) of certain policies that are quite close to the Nordhaus projections themselves.

Having access to a wide range of models proved, therefore, to be highly instructive to the MRG. Differences in model results stimulated further research to determine the causal factors behind those differences. Results of that research presented in this section highlight the importance of elasticity assumptions (own and cross) in interpreting model results.

IV AN EXPLORATORY ANALYSIS OF RESEARCH, DEVELOPMENT, AND DEMONSTRATION PROGRAMS UNDER UNCERTAINTY

IV.1 THE CONTEXT OF RESEARCH AND DEVELOPMENT OF ENERGY TECHNOLOGIES

All research and development of industrial processes finds its origin in basic research in the sciences, some of it performed in the pursuit of pure knowledge and some already motivated by the possibility of subsequent application. In these early stages, the characteristics of future technologies that may ultimately result are not sufficiently clear that a decision for further funding or support can be based on a quantitative cost-benefit analysis. Where the allocation of funds to such basic research is motivated by the possibility of applications, its amount depends mostly on the perceived social and national importance of obtaining increased knowledge in the area in question, as well as on the size and capabilities of the scientific talent in the field.

As work progresses through the basic and applied research phase to the development and demonstration phase, the technical characteristics become more clear and the costs of pursuing the research increase significantly. At this stage it becomes more meaningful to ask whether the direction proposed is promising and a good way to use the available resources. Quite often premature commitment to advanced development or demonstration projects results in a system that is a technological success but a market failure. At the same time such commitments tend to starve the more basic effort, and the basis for a more ramified evolution or development of the concept may be destroyed. Thus it is critical to evaluate the potential economic and social benefits of a technology at the point where a decision must be made on such a commitment.

Ideally, the assessment of potential costs and benefits should be performed on all technologies that compete for development and demonstration funds. It is also clear that the technology of advanced nuclear reactors, here examined as one of the focal points of CONAES, has received a good deal of study. In particular, the breeder reactor has been the subject of much cost-benefit analysis and is well down the development path. Nevertheless, a new decision point is upon us, and it is appropriate to analyze both the breeder reactor and alternative advanced reactors in a demonstration of the R&D decision-analysis methodology. A substantial amount of the more traditional cost-benefit analysis has been performed and has led to the formulation of the current development plan for the breeder. Many breeder and ACR concepts, as well as alternative fuel cycles, have been evaluated over the past 10 years by the Atomic Energy Commission and, more recently, by ERDA. The cost-benefit analyses that have led to the present program used a time-phased linear programming model of the electric sector, where electric demands were specified exogenously. The models employed by MRG represent a significant extension of that methodology in embedding the electric sector in the overall energy system and including the coupling of the energy system to the economy.

The concentration on the assessment of advanced reactors is based, first, on the importance of that technology to the CONAES study and, second, on the fact that most of the models employed gave some emphasis to the modeling of the alternative nuclear technologies that are either available now or under active development. Further, the R&D decision analysis approach employed in this chapter was developed specifically for the purpose of the economic evaluation of alternative advanced reactor development plans. As in the earlier sections of this report, the assessment of environmental and other risks is not included in the analysis of this section.

IV.2 OVERVIEW

As the work of MRG progressed, it became apparent that a general framework was required to structure the input assumptions and model results for application to the RD&D decision problems that CONAES and ERDA must deal with. To provide such a framework, the methodology of decision analysis was selected and has been applied to two problems.

The first application, described in Section IV.4, concerns the evaluation of alternative RD&D programs for the nuclear breeder reactor. This particular technology was selected as a sample case because of the availability to the group of earlier work along these lines. The methodology reflects the fact that RD&D decisions must be made at successive points in time. Uncertainties exist at any point in time; however, some may be resolved by the time that future decision points are reached. The methodology makes provision for direct incorporation of the uncertainties that exist at the various decision points.

The second application, described in Section IV.5, concerns choices among four alternatives: development of an FBR, of an ACR, of both, or of neither. For either reactor type there is a further choice between four alternative dates of availability. To accommodate this complexity of choices, the analysis compares only once-and-for-all decisions.

Because of the short time available for analysis and for estimation of the judgmental probabilities that describe the major uncertainties, the analyses presented here must be considered to be a preliminary demonstration of a methodology. Further analysis is needed to support the major decisions regarding the choices among alternative nuclear strategies.

The decision analysis approach is applicable to a special class of RD&D program evaluations where the R&D is directed towards technical and economic goals that can be quantified. It is recognized that these goals may not be fully fixed in advance and that success of the R&D toward these goals is not ensured; indeed, the methodology employs judgmental probability estimates of these important parameters. Decision analysis may be used to indicate the expected value of net benefits of the R&D given the uncertainties involved in the R&D program itself and in the future state of the world or competing systems. Conversely, the analysis also may be used as a basis for the definition of R&D objectives. In general, the methodology is more useful when a group of R&D programs is to be compared with respect to net expected value. In this demonstration application only one technology has been evaluated; therefore the results may not be interpreted in terms of relative R&D priorities in different technical areas.

At the present state of the art, the methodology is not appropriate to the evaluation of basic research programs where the objective is increased scientific knowledge. Even in applications involving an advanced development or demonstration project, care must be taken that all benefits and costs are properly accounted for. In some instances, R&D on one technology can have "fallout" benefits that accrue to other programs or markets; on the other hand, certain federal programs may inhibit the private sector from participating in an effective way. To the extent that such effects may be quantified, such considerations should be incorporated in the analysis. In any event, the results of the decision analysis are but one input to any RD&D decision problem and other aspects of the decision, whether quantifiable or not, must be given careful consideration. The application of decision analysis presented here is limited to a selected set of factors that can be incorporated directly into the methodology.

The work of the MRG produced quantitative estimates assessing the role and impact of specific energy technologies on the future energy system of the United States. The results of the modeling work can thus provide major inputs to decision analysis methodologies such as cost-benefit analysis. Indeed, cost-benefit analysis has been applied to the breeder reactor program in several studies performed prior to the CONAES study.

Most cost-benefit analyses that have been performed on energy technologies were seriously deficient in their treatment of uncertainty. It is obvious that the ultimate economic benefits of a new technology are heavily dependent on such parameters as the size of the market that develops for the form of energy, e.g., gas, liquid, or electric; the cost and characteristics of competing technologies; and the outcome of the RD&D program for the new technology. Furthermore, in addition to the strictly economic considerations, the potential social and environmental impacts of large-scale commitments to alternative technological systems must be carefully weighed. There are great uncertainties with respect to all of these parameters that must be made explicit and dealt with in making a decision on the appropriate budget level and priority of a particular program.

The usual approach to dealing with uncertainties in a cost-benefit analysis involves the definition of alternative deterministic scenarios that bound the anticipated range of variability. This procedure was applied in Chapter III. One of its limitations is that it generates numerous cases that usually show different benefits with a case for every major viewpoint. In addition, each analysis is performed assuming that the technology in question is developed, demonstrated, commercialized, and implemented in the energy system and that the immediate decision deals with the full commitment to the technology through all phases from development through implementation. Clearly, if the benefits of the technology are insensitive to the outcome of all of the uncertain parameters within their estimated range of variability, the decision is straightforward. The usual case, however, is such that there are significant variations in benefits with different assumptions, and that better information will be available at future dates when decisions must be made either to terminate a project or continue it through the next phase. Thus, it is necessary to structure the decision problem so as to reflect judgments of the likelihood of various outcomes of critical parameters as well as the sequential nature of the decision problem. In this way the current decision to be faced deals with the continuation of a program only through an interval of time to a point where better information on major uncertainties can be developed or will become apparent.

The analysis of this chapter improves on that of Chapter III by explicitly introducing measures (judgmental probabilities) of uncertainties that enter into the problem. As said above, it still shares with the earlier sections the defect that environmental and other noneconomic policy considerations, and the uncertainties connected with these, are not taken into account.

The previous work of Richard G. Richels* provided an example of the application of decision analysis to the evaluation of alternative R&D strategies under conditions of uncertainty. This work was performed under the guidance of Alan Manne and involved the study of the U.S. breeder reactor program. The model ETA was used in that study. Because of the direct applicability of Richels' work to R&D decision analysis, the R&D Sub-group decided to use it as a starting point and extend the work to incorporate the major uncertainties that had been dealt with in the modeling work performed by the MRG. The input parameters used for the MRG modeling runs were reviewed and modified where appropriate on the basis of the previous MRG work. New information has also been obtained from other panels and groups, and the new round of R&D analysis runs provided an opportunity to include some of these new data. The R&D analysis addressed the following alternative breeder strategies.

1. Concurrent development of Clinch River and the commercial prototype.
2. Sequential development of Clinch River and the commercial prototype.

*"R&D Under Uncertainty: A Study of the U.S. Breeder Reactor Program," Energy and Environmental Policy Center, Harvard University, 1976.

3. Wait for more information on important parameters but continue R&D at a reduced level.
4. Stop R&D altogether.

Another alternative strategy that has been considered involves international cooperation on the development of the breeder. The timing of the program elements under this strategy was assumed to be similar to that for the concurrent development option.

The operational procedure followed for the above decision options involved the definition of major decision points in time and of uncertainties that are likely to be resolved in the various time frames, the estimation of judgmental probabilities for the outcomes of each of the major uncertainties, and the calculation of costs and benefits for alternate paths in the decision tree defining specific combinations of decisions and outcomes of uncertain events.

The decision options and outcomes of uncertain events were represented in a decision tree that shows the alternative combinations that may occur over time. Judgmental probability estimates were made using a survey approach among MRG members and consultants. In a complete policy study using this methodology, specialists should be consulted concerning each individual area of uncertainty. Although this was not possible in this study for each critical parameter, two members of MRG had the opportunity to meet with CONAES' Uranium Resource Group (URG) and elicit that group's estimates regarding uranium resources subsequent to the initial analysis. As an alternative to the decision analysis based entirely on the responses of MRG members and consultants, an alternative case was analyzed in which the probability estimates relating to uranium availability were replaced by the URG's probability estimates.

Because of the limited set of inputs to the estimates, it must be stressed again that the analysis performed is to be considered as a demonstration of the capability of the decision analysis methodology. Insights into the decision problem may be obtained; however, the user is cautioned to evaluate carefully the decision structure and the definitions and probability estimates of the uncertain events.

In Section III.8, the economic value of a technology was estimated by a sum of discounted economic benefits minus costs associated with the use of that technology. In this chapter, the same concept is applied to evaluate any future path of the energy economy that arises from a given sequence of outcomes of uncertain events as revealed, combined with a given sequence of decisions taken when due. At each point in time where a choice between alternative decisions is due, the economic value of each alternative is calculated as a probability-weighted sum of the economic values of all the then future paths that the roll of the dice may open up--provided that at each subsequent point the decision then having the highest value is chosen.

The exposition of Section IV.4 will show that this procedure is *not* circular. It is mentioned here only to bring out one limitation of the method. A "probability-weighted sum," also called an "expected value" or a "mean value" of a set of outcomes, will treat as equally desirable (1) a fully predictable value, (2) a 50-50 chance between a gain of $5 billion and a loss of $5 billion relative to that predictable value, and (3) a 10-to-1 chance for a gain of $10 billion against a loss of $100 billion.

Clearly, there will be many who would not be quiescent in the face of such a large potential loss, even though the loss may be highly unlikely. Put another way, individuals and society may be averse to the risk of large, low-probability losses, and may wish to purchase "insurance" to protect against such risks. The decision analysis calculations reported here do not adjust for the existence of risk aversion. How society wishes to behave in the face of such risks is a choice the MRG has not tried to predict, or even to recommend. Nevertheless, a final decision on breeder or any other advanced nuclear technology should weigh the expected values of net discounted economic benefits against the economic, social, and environmental risks involved. Expected value calculations, taken alone, tend to concentrate scarce R&D resources in one or a few technologies and to lead away from investments in a larger diversity of programs designed to ensure against adverse outcomes among the few. At the same time, however, such estimates provide highly important information to be used in conjunction with social attitudes toward risk, in the evaluation of the strictly economic merits of a particular R&D program.

In Section IV.4, the sum of discounted economic benefits minus costs of breeder reactor technologies is estimated for various branches of the decision tree using the ETA model. This information, together with estimates of the discounted cost of RD&D, provides the costs and benefits of various combinations of decisions and uncertain events expressed in 1975 dollar equivalents. The costs and benefits of each path are then multiplied by the judgmental probability estimates and added to give the expected economic value (or cost) of specific decision options. The results presented were obtained using the R&D Sub-group probability estimates of MRG members and consultants; however, sensitivity analyses may be performed quite easily for other sets of estimates.

Earlier work of the MRG presented in Chapter III showed that an ACR with high conversion ratio (~0.8) had significant economic benefits over the mid to high range of uranium availability and displaced the breeder reactor under those circumstances. Because of the limited time available, it was decided that the R&D analysis runs with ETA would be performed without an advanced converter. The results of the ETA runs are summarized in Section IV.4.

The Nordhaus model was also applied to the R&D decision analysis for the calculation of benefits for different outcome branches of the decision tree. It was decided that the series of runs performed with this model would focus more directly on the competition between the advanced converter and breeder. The results of the Nordhaus analysis are presented later in Section IV.5.

Despite the preliminary nature of this demonstration of decision analysis, it is possible to draw some general conclusions regarding the decision problem of alternative breeder and/or ACR development strategies. Further, the analysis indicates which parameters bear most directly on the decision and thus gives guidance towards areas where expert judgment should be developed.

Using the judgmental probabilities estimated by MRG members and consultants for all variables, the ETA analysis in Section IV.4 indicated that the economically favored breeder reactor strategy was to stop R&D. Using the alternative estimate of uranium resources developed through discussion with the URG, the optimal strategy becomes concurrent development of the Clinch River reactor and the commercial prototype. The expected value of the net benefits is $4.5 billion in this latter case. The analysis of Section IV.5 with the Nordhaus model indicates that the net benefits of the breeder reactor are near zero and that those benefits are sensitive to both the timing of the R&D and the availability of an advanced converter. Positive net benefits, in the range of $3 billion to $6 billion, result under circumstances where the entire breeder R&D program, including the commercialization phase, is deferred by 10 to 20 years and where the advanced converter is not available. If an advanced converter is available by the year 2000, the analysis with the Nordhaus model indicates low breeder net economic benefits.

In addition to estimating the expected value of net benefits of nuclear reactors, both ETA and Nordhaus have estimated the "standard deviation" of the value. It is important to understand what the standard deviation in these calculations represents. Specifically, each standard deviation reported refers to one specific policy. It holds that policy constant through many alternative future courses of the world that differ according to the roll of nature's dice in the particular history. The standard deviation registers the dispersion of the probability distribution so generated. In several cases, this standard deviation is large compared with the expected net benefits for the policy in question. It must therefore be emphasized that the standard deviation should *not* be interpreted as a "standard error of estimate," or measurement error that might seem to swamp the estimated expected benefit. Policy choices are, by their nature, choices between two or more courses of action. For any given pair of policies, the roll of some dice will affect the net benefit in much the same way, whereas the rolls of the other dice will favor the outcome of one policy over that of the other. By and large, therefore, we expect that the standard deviation of the difference in the net benefits from two alternative policies is substantially smaller than that for the outcome of either policy by itself. It is the latter that is recorded in Sections IV.4 and IV.5.

The ETA and Nordhaus analyses turn out to be quite consistent in their estimates of the standard deviation of expected benefits based on the MRG probability estimates. The standard deviation comes out, in both cases, at about $20 billion. The large standard deviation indicates that, whatever the estimated expected value, the actual realization will possibly be considerably larger or smaller.

The breeder development option of cooperation with other nations was evaluated and is estimated to have additional benefits roughly equal to any reduction in the cost of the program to the United States that may result from such cooperation. This, of course, assumes that the technical characteristics of the resulting plant are identical for the two strategies. However, other factors, such as the possible loss of overseas markets, should also be considered in evaluating the cooperative R&D options.

Sensitivity analyses of the results of both Sections IV.4 and IV.5, as well as differences in certain assumptions that were made in the two models, indicate areas where further work is needed. Among these are:

1. *Characterization of the advanced converter in terms of RD&D costs for the plant and fuel cycle, and of technical characteristics including conversion ratio and possible process heat applications.* The Nordhaus analysis indicated that an ACR was closely competitive with the breeder, primarily due to lower RD&D costs. Therefore, additional study should be made of the R&D costs.

2. *Definition of breeder R&D costs, separating supporting research from major demonstration projects. Consideration of alternative fuel cycles for the breeder reactor.* The cost of commercialization for the first 6 GW of capacity also needs to be estimated. In the ETA analysis it was assumed that the cost of commercialization was about $2 billion undiscounted, while the Nordhaus analyses assume a cost of roughly $6 billion undiscounted (approximately 1,000 $/kW for the first 6 GW of commercial capacity). Another aspect of the R&D costs that must be considered is the effect of stretch-out. The ETA probabilistic analysis assumes that, for later implementation dates, the R&D effort is simply stretched out, while the Nordhaus analysis assumes that the R&D program is carried out in the last 10 years prior to commercialization phase in all cases.

3. *Judgmental probability estimates of experts.* As indicated previously, it is recommended that probability estimates be developed by appropriate experts before any further analyses are undertaken. Sensitivity analyses presented in Section IV.3 indicate that the critical questions associated with the probability estimates deal with:

- uranium resource availability; there is a need to reconcile the geologists' viewpoint of limited supply at intermediate costs with the economists' view of a rising supply curve with allowance for the short history of the industry, for different modes and depths of occurrence and different methods of discovery and extraction;
- public acceptance of plutonium fuel cycles;
- ability to produce large quantities of coal (>50 Quads) at a reasonable direct and environmental cost; and
- anticipated electric energy demand growth.

4. *Resolution of key uncertainties.* The two analyses were based on different assumptions regarding the date of resolution of uncertainties. In the probabilistic analysis of ETA results, it was assumed that most uncertainties would be resolved in 1986, with the result that R&D and commercialization expenditures past that point could be avoided if the resolution of uncertainties indicated no breeder benefits. One of the uncertainties that is assumed to be resolved by 1986, for example, is the cost of the Alternative Energy System (AES). In several models, AES's are not defined explicitly and represent a set of possible technologies that have been construed to include solar heating, conservation, new discoveries of low-cost oil, and fusion. Clearly, the time frames of

development for this broad mix of systems cover a wide range and not all will be defined by 1986. It is likely, however, that in view of the current policy encouraging conservation and solar demonstrations that much will be known of the characteristics of these systems by 1986. In further work, it would be best to divide AES's between near-term technologies and long-term technologies.

In contrast with the analysis of ETA results, the Nordhaus analysis assumes that uncertainties will not be resolved until after the commercialization phase. These different assumptions have a significant effect on the net benefits of alternative decisions and should be resolved in any future work.

5. *Independence of probability estimates.* In the R&D analysis, it has been assumed that the probability estimates for the outcomes of different variables are independent. This is clearly a convenience for analysis and is an assumption that must be carefully evaluated. In particular, it may be that the question of public acceptance of the breeder and plutonium fuel cycle is closely related to its anticipated economic benefits as well as to its perception of the various risks involved and, thus, to the other uncertain parameters that bear on those benefits and risks. If, as an extreme case, the question of public acceptance were assumed to be perfectly correlated with anticipated economic breeder benefits, then this parameter would be effectively removed from the calculation and, as indicated in Table IV.6, economic breeder benefits would increase by about $8.1 billion (from $4.5 billion to $12.6 billion).

Further, the probability estimates have been assumed to be independent of the decision. It can be postulated that the decision itself and resultant policies will influence the outcome of some uncertainties. For example, it is conceivable that a decision to forego the breeder option would result in price increases for uranium. To the extent that coal is a close substitute for the generation of electricity, some competitive constraints would apply; however, suppose the coal option were also blocked or limited by environmental problems. Such interrelationships and feedback effects between the decision options and the probability estimates of uncertain events should be explored in future work.

6. *Consideration of the effects on the environment, on national and world security, and on related social and public policy concerns.* As in the interpretation of the work described in Chapter III, the present work emphasizes the economic benefits, costs, and risks, leaving aside the "environmental" effects in the broadest sense of the word. Quantification of these considerations is needed wherever possible; otherwise, judgmental assessment of the importance of these considerations is necessary for a judicious use of the techniques here presented.

Granting the importance of these observations, it is suggested that the decision analysis methodology applied by the R&D Sub-group is a natural way to structure complex decision problems. The key step is the development of judgmental probability estimates; therefore, it is imperative that careful attention be given to the estimation of uncertainties by a balanced selection of technical experts qualified in specific areas. Decision analysis thus serves to provide a quantitative assessment and display of the range of uncertainties and the effect of those uncertainties on the decision problem. It does not overcome all of the difficulties associated with deterministic analyses in that polarization over alternative scenarios in such analyses may simply be replaced by polarization over alternative probability distributions that describe critical uncertainties.

The methodology can be extended to other R&D programs. Being based on the quantitative analysis of benefits and uncertainties, it is an approach that R&D personnel would find appropriate for decision problems. In view of the fact that one of the more useful characteristics of the methodology is the structure it provides for evaluating alternative decision options, it has considerable promise for the analysis of policy options as well as technical options.

IV.3 DETERMINATION OF JUDGMENTAL PROBABILITY ESTIMATES

The R&D decision analysis method employed by the R&D Subgroup required the use of probability estimates regarding the resolution of a variety of uncertainties. A survey technique was employed to provide values for these probability estimates. In this section, the technique employed, the questions asked, and the resulting judgmental probability estimates are discussed.

The technique employed to determine the probability estimates was based on the polling of a group of experts who were assumed to be knowledgeable in the subject areas of interest. In this technique, consecutive rounds of a questionnaire covering the material at issue are circulated and it is hoped that a consensus of the group can be achieved as the individual respondents become familiar with the estimates and reasoning of their colleagues. For the decision tree probability estimates, the "experts" consisted of the participants in the MRG and consultants to that group, including economists, energy policy analysts, and energy technology analysts. Two consecutive polls were conducted of the group using essentially the same questions, with the members being provided with the statistical results of the first poll before the second round. Consultation and discussion of the questions between rounds was on an informal basis. There were six first-round responses. The nine second-round responses were employed to determine the probability estimates used in the R&D analysis.

The survey questions as incorporated in the second (final) round of the survey and the respondents to the final round are listed in Appendix F.1. Note that questions 2 and 3b relate to policy variables, while all others refer to realization variables.

The results of the analysis of the second-round responses are shown in Table IV.1. For each value of a variable, the mean, range of responses (lowest and highest estimates), and standard deviation of the nine estimates are shown. In addition, where applicable, the expected value of the variable outcome is computed using the mean of the estimates for the separate values (i.e., $E(x) = \Sigma p_i x_i$ where $E(x)$ is the expected value of the variable x and p_i is the probability of the outcome x_i).

The following points are noted regarding the responses:

1. In the case of variables for which the MRG has chosen base values, the expected values computed from the judgmental probability estimates are reasonably close to the base values.

2. The range in the individual responses (for the probability estimates) is large, and the standard deviations are mostly about one-half to two-thirds of the value of the mean. These observations indicate that a close consensus was not achieved among the respondents after two rounds. Accordingly, the mean judgmental probability estimates as incorporated in the R&D decision tree should not be interpreted as providing a precise description of the uncertainties.

IV.4 DECISION ANALYSIS OF BREEDER DEVELOPMENT USING ECONOMIC BENEFITS CALCULATED BY THE ETA MODEL

Calculation of Breeder Benefits

Under the MRG's Base-Case assumptions, the economic benefits of the LMFBR are virtually zero. This can be seen directly from the levelized cost comparison in Table III.11. Moreover, this same result is obtained by scenarios that incorporate the effects of resource depletion and rising uranium costs. (See Table III.20, especially the comparisons between the objective function for MRG scenarios 1, 2, 4, and 5.)

ETA model results were used in the R&D decision analysis for selected combinations of outcomes of uncertain events as described previously. The first step involved the calculation of breeder benefits for 10 combinations of outcomes. The benefits calculated for each combination were then used in the probabilistic analysis to determine the expected net benefits of alternative decision options.

Throughout the R&D analysis, a number of key assumptions that were adopted by the MRG were retained. The major ones that have been retained but that may warrant further scrutiny are: (a) a "real" pretax rate of return of 13 percent for pricing and investment decisions within the electric utilities industry; (b) a "real" posttax rate of return of 6 percent for public R&D decisions; (c) the specific assumptions on investment costs per unit of capacity, the plant capacity factors, and the fuel costs assumed for coal-fired, nuclear, and solar-electric units; and (d) the rates and dates of introduction postulated for new energy-supply technologies.

The date of introduction of the LMFBR is defined as the first year in which 6 GW of demonstration and prototype capacity become available. Three alternatives have been considered with respect to this date: the year 2000, 2010, and 2050 or later. This last case (in effect, no breeder program) serves as a bench mark against which breeder benefits may be measured. Economic benefits are defined as the difference in the economy's total discounted energy costs--with versus without the breeder. In the ETA model, this comparison includes both the costs of energy supply and also the costs of price-induced conservation and interfuel substitution. Demands will vary from one case to another, depending upon the prices of electric and nonelectric energy.

TABLE IV.1 Judgmental Probability Estimates[a]

Variable (x)				Probability Estimate			
				Mean	Range	$\sigma(x)$[b]	$E(x)$[c]
1(a)	Energy growth rate (average 1975-2000):[d]	1.14%/ year		0.36	0.049-0.8	0.28	
		3.1	B[e]	0.52	0.2 -0.95	0.24	2.5%
		4.7		0.11	0 -0.45	0.14	
(b)	Total non-price induced conservation by 2010:	10%	B[e]	0.62	0.2 -0.8	0.19	
		None		0.38	0.2 -0.8	0.19	6.2%
(c)	Price elasticity:						
	(i) Electricity	0.1		0.13	0 -0.3	0.10	
		0.3		0.35	0.09 -0.6	0.17	0.43
		0.6		0.52	0.1 -0.9	0.24	
	(ii) Nonelectric fuels	0.1		0.16	0 -0.4	0.11	
		0.3		0.38	0.2 -0.6	0.15	0.41
		0.6		0.46	0.05 -0.7	0.23	
(d)	Population growth rate:	0.5%/ year		0.23	0.1 -0.4	0.09	
		0.8		0.58	0.5 -0.8	0.09	0.81
		1.2		0.19	0.1 -0.3	0.08	
2	Coal/shale constraints:	None	B[e]	0.38	0.1 -0.8	0.26	
		No more than 50 Quads		0.62	0.2 -0.9	0.26	
3(a)	Cost of non-electric alternative energy system:	2$/$10^6$ BTU		0.18	0.02 -0.5	0.13	
		5	B[e]	0.56	0.3 -0.8	0.16	5.3 $/$10^6$ BTU
		8		0.27	0 -0.6	0.18	
(b)	Oil imports: As percent of oil and gas consumption:	10%		0.23	0.03 -0.45	0.15	
		33%		0.54	0.4 -0.67	0.10	36%
		70%		0.23	0.1 -0.5	0.12	
4	Public acceptance:	No nuclear		0.15	0.03 -0.5	0.14	
		LWR only		0.15	0.005-0.3	0.09	
		ACR & LWR		0.16	0.047-2.5	0.07	
		LMFBR & LWR		0.14	0 -0.5	0.15	
		LMFBR & ACR	B[e]	0.40	0.1 -0.903	0.27	

TABLE IV.1 Continued

Variable (x)				Probability Estimate Mean	Range	$\sigma(x)$[b]	$E(x)$[c]
5	Uranium resources:[e]	Low		0.23	0.015-0.5	0.17	
		Intermediate	B[e]	0.41	0.015-0.7	0.20	
		High		0.37	0.2 -0.97	0.25	
6	LMFBR capital cost:	1 × LWR[f] cost		0.10	0 -0.2	0.07	
		1.25 × LWR	B[e]	0.56	0.4 -0.7	0.11	31%
		1.5 × LWR		0.34	0.2 -0.6	0.14	
7	Advanced converter (ACR) capital cost:	1 × LWR[f]		0.15	0 -0.3	0.09	
		1.1 × LWR	B[e]	0.51	0.4 -0.6	0.07	13%
		1.25 × LWR		0.33	0.1 -0.6	0.14	
8(a)	Solar-electric capital cost:	1.25 × LWR[f] cost		0.22	0.05 -0.6	0.16	
		2 × LWR		0.29	0.15 -0.4	0.08	>2.7
		2.6 × LWR	B[e]	0.24	0 -0.4	0.11	
		>5 × LWR		0.24	0 -0.5	0.16	
(b)	Fusion capital cost:	1 × LWR[f] cost		0.25	0.05 -0.35	0.09	
		2 × LWR		0.29	0.05 -0.45	0.12	
		3 × LWR		0.21	0.05 -0.3	0.09	
		5 × LWR		0.16	0.05 -0.2	0.07	
		∞		0.21	0.1 -0.4	0.10	

[a]Results of second-round poll--nine responses. See questionnaire, Appendix F.1, for a comprehensive statement of questions.
[b]Standard deviation.
[c]Expected value.
[d]Definition of low-, intermediate-, and high-energy demand growth and uranium resource levels as used in the R&D analysis were revised after the questionnaire was circulated. Using the mean judgmental probability estimates from this table and the original resource values from the questionnaire, a cumulative probability distribution curve was drawn from which probability estimates for the revised resource levels were derived.
[e]B = Base assumption.
[f]LWR capital cost (excluding initial fuel inventory) assumed to be $650/kW.

TABLE IV.2 States-of-the-World with Potential Economic Benefits for the LMFBR[a]

Realization Leading to LMFBR Benefits That Are:	Energy Demand Growth Rate at Constant Prices, 1970-2010	Nonelectric AES Cost ($/Million BTU)	Uranium Resources (Million Short Tons at $45/lb)	Objective Function, Present Value of the LMFBR's Gross Economic Benefits ($ Billions, Discounted at 6 Percent) when LMFBR Introduction Date Is:		
High (H)	3.4%	$8	1.6 (+ 0.5 at $75 + 0.15 at $120)			
Moderate (M)	2.9%[b]	$5[b]	3.7 (+ no additional quantities)			
Low (L)	1.5% (approximately zero after 2010)	$2	3.7 (+ additional quantities at higher prices)[b]	2000	2010	2050 or Later
	H	H	H	139	48	(7,523)[c]
	H	H	M	136	53	(7,402)[c]
	H	M	H	97	50	(6,934)[c]
	H	M	M	70	44	(6,853)[c]
	M	H	H	123	49	(6,178)[c]
	M	H	M	76	52	(6,086)[c]
	M	M	H	59	36	(5,853)[c]
	M	M	M	26	24	(5,795)[c]
	L	H	H	0	0	(4,194)[c]
	L	M	M	0	0	(4,172)[c]

[a]Assumptions identical to the MRG runs except for the following changes: (i) coal limits of 50 Quads are imposed, (ii) the advanced converter reactor is excluded, and (iii) the LWR option is included both with and without plutonium recycle. By taking the difference between the objective function value for MRG scenario no. 5 and the corresponding run here for the LMFBR delayed until 2050 or later, we can calculate the gross economic benefits of plutonium recycle into the LWR. This amounts to $17 billion, discounting at 6 percent.

[b]The low case defined here is the MRG Base Case.

[c]This is the optimal value of the objective function (the sum of producers' and consumers' surpluses). It is a meaningless number except as a reference point for those scenarios in which the breeder becomes available either in the year 2000 or in 2010.

Table IV.2 summarizes the results for states-of-the-world, resulting in positive breeder benefits for each of the three possible introduction dates. Altogether, 30 computer runs were performed in order to evaluate breeder benefits under 10 alternative states-of-the-world.* These states are defined through combined variations in the effects of three key variables: (a) the rate of growth of energy demand under constant prices from 1970 through 2010, (b) the cost of AES's to replace our limited resources of oil and gas, and (c) the cumulative quantity of uranium resources available. Breeder benefits will depend upon the eventual "realization" of these uncertain parameters.

Each state-of-the-world is defined through a triplet of letters. For example, MHH refers to that state in which the rate of energy demand growth is *medium* (i.e., the MRG's Base Case with respect to GNP growth and nonprice conservation); in which high costs of AES's ($8 per million BTU) are experienced tending toward *higher* breeder benefits; and the uranium resources realization (1.6 million short tons available at $45 per pound) is low, tending toward *higher* breeder benefits. For this case, Table IV.2 indicates that breeder benefits would be $123 billion if this technology is introduced in the year 2000, but that benefits would drop to $49 billion if there is a delay until 2010.

To elucidate the nature of the results obtained from the ETA model, detailed results for the MHH case are presented--first for breeder introduction in the year 2000 (Table IV.3) and then for the case of delay until the year 2050 or later (Table IV.4). According to Tables IV.3 and IV.4, with the MHH state-of-the-world, it would be economically optimal to introduce the breeder at the maximum rate postulated by the R&D Subgroup, but because of the high costs of solar electricity assumed in this scenario, it would *not* be optimal to introduce this latter technology at the maximum rate.

For comparability with other MRG calculations, two caveats should be noted:

1. Here the highest demand growth rate is considerably lower than that in the MRG's scenario no. 7. According to Table IV.2, "high" means a 3.4 percent rate of energy growth at constant prices from 1970 through 2010. The 3.4 percent rate is chosen so that the total demand in 2010 would be 10 percent above that which would be projected from a constant energy-GNP ratio and the MRG's *Base-Case* GNP growth rate.
2. For assessing R&D decisions, breeder benefits have been calculated on the assumption that it will be acceptable to recycle plutonium into LWR's. If we rule out plutonium recycling into LWR's, it is estimated that the additional cost over that of MRG scenario no. 5 would be $17 billion (see footnote *a* in Table IV.2).

Aside from the two low-demand cases (LHH and LMM), Table IV.2 contains only those cases in which there were prior reasons to believe that breeder benefits would be significant. Previous runs had already

*Copies of the complete output results can be obtained from Alan Manne upon request.

TABLE IV.3 Scenario: MHH; FBR Introduced in Year 2000

	1975	1985	1990	2000	2010	2020	2030
4. Domestic energy production, by source, net of exports (Quads)	(57.9)	(70.9)	(80.5)	(109.5)	(133.8)	(164.3)	(206.6)
4.1 Coal, total:	(13.3)	(17.8)	(20.1)	(40.0)	(45.0)	(47.7)	(49.0)
a. for electricity generation	8.8	12.3	14.0	17.6	18.8	14.2	6.7
b. for synthetic fuels	---	---	0	15.0	17.2	22.5	28.9
c. direct use and other	4.5	5.5	6.1	7.4	9.0	11.0	13.4
4.2 + 4.3 Natural gas, oil, and NGL	39.7	42.0	44.2	38.1	28.8	13.2	5.0
4.4 Nuclear, total:	(1.7)	(7.1)	(11.6)	(23.5)	(37.6)	(50.7)	(82.7)
LWR	1.7	7.1	11.6	23.1	30.8	23.4	11.8
FBR	---	---	---	0.3[a]	6.8[a]	27.3[a]	70.9
Other (HWR, HTGR)	---	---	---	0	0	0	0
4.5 Solar electric	---	---	---	0	0	0	0
4.7 Shale oil	---	---	---	2.0	6.0	10.0	11.5
4.8 + 4.9 Hydro, geothermal, etc.	3.2	4.0	4.6	5.9	7.5	9.6	12.3
4.10 Nonelectric AES	---	0	0	0	8.9	33.1	46.0
5. Oil and gas imports (Quads)	13.2	21.0	22.1	19.1	14.4	6.6	2.5
6. Total energy consumption (Quads)	71.1	91.9	102.6	128.6	148.2	171.0	209.1
7. Electricity generation (trillion kWh)	2.0	2.9	3.5	4.8	6.5	7.5	10.2
7.1 Percent oil and gas	32	18	11	---	---	---	---
7.2 Percent coal	44	44	42	39	30	20	7
7.3 Percent nuclear	8	25	33	49	58	67	81
7.4 Percent other	16	14	13	12	12	13	12

8. Uranium consumption (million tons of U_3O_8, cumulative from 1975)	---	0.2	0.4	0.8	1.5	2.0	2.3
9. Domestic prices:							
9.1 Coal ($/$10^6$ BTU)	---	0.7	0.7	0.8	3.5	3.5	3.5
9.2 + 9.3 Oil and gas ($/$10^6$ BTU)	---	2.9	3.2	4.1	8.0	8.0	8.0
9.4 Electricity ($/1,000 kWh)	---	22.5	23.1	25.2	37.4	38.0	31.0
9.5 Uranium ($/lb U_3O_8)	---	83.9	97.1	138.3	212.2	334.2	524.0
10. Cumulative undiscounted investment (billions of 1975 dollars, cumulative from 1975)							
a. Total energy sector	0	457	718	1,357	2,283	3,556	5,094
b. Electricity generation, transmission, and distribution	0	187	318	661	1,117	1,572	2,505

[a]Upper bound on rate of FBR introduction.

NOTE: Due to rounding, totals need not add.

TABLE IV.4 Scenario: MHH; FBR Introduced in Year 2050 or Later

	1975	1985	1990	2000	2010	2020	2030
4. Domestic energy production, by source, net of exports (Quads)	(57.9)	(70.4)	(79.9)	(106.9)	(132.7)	(160.9)	(192.2)
4.1 Coal, total:	(13.3)	(19.9)	(22.9)	(40.0)	(45.0)	(47.7)	(49.0)
a. for electricity generation	8.8	14.4	16.8	20.0	19.2	13.5	6.4
b. for synthetic fuels	---	---	0	12.6	16.8	23.2	29.2
c. direct use and other	4.5	5.5	6.1	7.4	9.0	11.0	13.4
4.2 + 4.3 Natural gas, oil, and NGL	39.7	42.1	44.3	39.7	28.3	13.0	4.9
4.4 Nuclear, total:	(1.7)	(4.3)	(8.1)	(19.3)	(35.5)	(32.7)	(21.5)
LWR	1.7	4.3	8.1	19.3	35.5	32.7	21.5
FBR	---	---	---	---	---	---	---
Other (HWR, HTGR)	---	---	---	---	---	---	---
4.5 Solar electric	---	---	---	0	0	13.0	40.6
4.7 Shale oil	---	---	---	2.0	6.0	10.0	11.5
4.8 + 4.9 Hydro, geothermal, etc.	3.2	4.0	4.6	5.9	7.5	9.6	12.3
4.10 Nonelectric AES	---	0	0	0	10.4	34.9	52.3
5. Oil and gas imports (Quads)	13.2	21.1	22.1	19.8	14.2	6.5	2.4
6. Total energy consumption (Quads)	71.1	91.4	102.1	126.7	146.9	167.4	194.6
7. Electricity generation (trillion kWh)	2.0	2.8	3.4	4.6	6.3	6.9	8.1
7.1 Percent oil and gas	32	18	11	---	---	---	---
7.2 Percent coal	44	53	51	46	32	20	8
7.3 Percent nuclear	8	15	24	42	56	47	26
7.4 Percent other	16	14	13	13	12	33	65

8. Uranium consumption (million tons of U_3O_8, cumulative from 1975)	---	0.1	0.3	0.7	1.3	1.9	2.2
9. Domestic prices:							
9.1 Coal ($/$10^6$ BTU)	---	0.7	0.7	1.0	3.5	3.5	3.5
9.2 + 9.3 Oil and gas ($/$10^6$ BTU)	---	2.9	3.2	4.1	8.0	8.0	8.0
9.4 Electricity ($/1,000 kWh)	---	23.7	23.9	26.6	38.9	42.5	42.5
9.5 Uranium ($/lb U_3O_8)	---	103.0	122.6	184.0	293.9	480.6	801.5
10. Cumulative undiscounted investment (billions of 1975 dollars, cumulative from 1975)							
a. Total energy sector	0	443	703	1,324	2,269	3,666	5,416
b. Electricity generation, transmission, and distribution	0	171	302	629	1,061	1,622	2,655

NOTE: Due to rounding, totals need not add.

demonstrated that benefits would be negligible or negative under any *one* of the following contingencies: (i) energy demand growth rates are low; (ii) there are no environmental limits on coal production and consumption; (iii) low-cost nonelectric AES's become available; (iv) ample amounts of uranium resources can be mined at low prices; (v) an ACR and fuel cycle become available (e.g., an HTGR with an 82 percent conversion ratio); or (vi) low-cost solar-electric or fusion power becomes economically competitive shortly after the year 2000. Any of these contingencies *could* occur, and this is the reason for a probabilistic analysis of R&D decisions. There is no reason, however, to calculate detailed supply-demand balances for these scenarios since, for the purpose of benefit-cost analysis, all that matters is that each of these circumstances would lead to the identical overall result: no economic benefits from development of the LMFBR.

Before concluding this section, one methodological issue ought to be mentioned. Investment decisions for the initial time periods (1980-1990) must be made *prior* to learning which of the future states-of-the-world will be realized. An error is introduced to the extent that the electric utilities' investment decisions for 1980-1990 would *not* be invariant with respect to uncertainties concerning the post-2000 world as has been assumed in this work. This problem can be avoided by using stochastic programming techniques rather than constructing 30 individual deterministic scenarios. The issue deserves further analysis and research. For present purposes, however, there is little reason to believe that this approximation creates serious errors.* For R&D expenditures, the timing of information is of course vital. This is why we now turn to a decision tree analysis of these expenditures.

Definition of the Decision Program

The starting point for the decision analysis involves graphing the problem using a decision-flow diagram, as a series of decision nodes occurring sequentially over time. Figure IV.1 shows the four alternative breeder investment options listed in Section IV.2: (1) concurrent development of Clinch River and the commercial prototype, (2) sequential development of Clinch River and the commercial prototype, (3) wait until better information is available (but continue R&D at a reduced funding level), and (4) stop the R&D program altogether. The option of a cooperative R&D program with other nations is assumed to be equivalent in terms of schedule, technical outcome, and net cost to the concurrent development option.

If one of the first three options is selected, additional decision points will be faced in the future. Figure IV.2 presents the entire decision-flow diagram. The squares denote decision points. Note that only one path leads to the first Commercial Breeder Reactor (CBR-1)

*This was the reason for the low value of perfect information found by A. S. Manne in "Waiting for the Breeder," *Review of Economic Studies Symposium*, 1974.

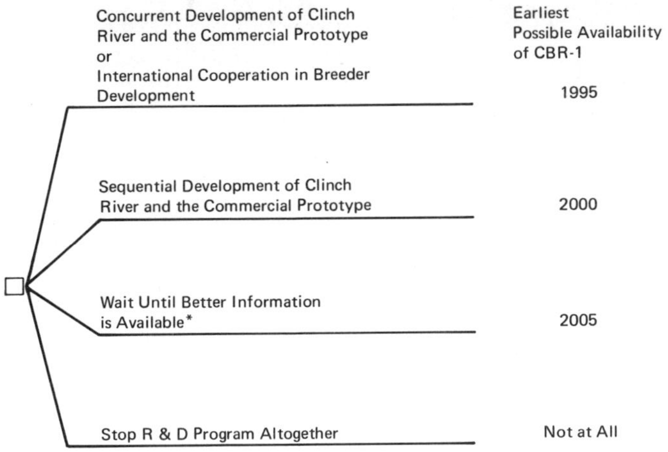

FIGURE IV.1 Different program options as of now (1977). This strategy involves a delay of the demonstration phase of the R&D program, but calls for continuation of basic research. See Appendix F.2.

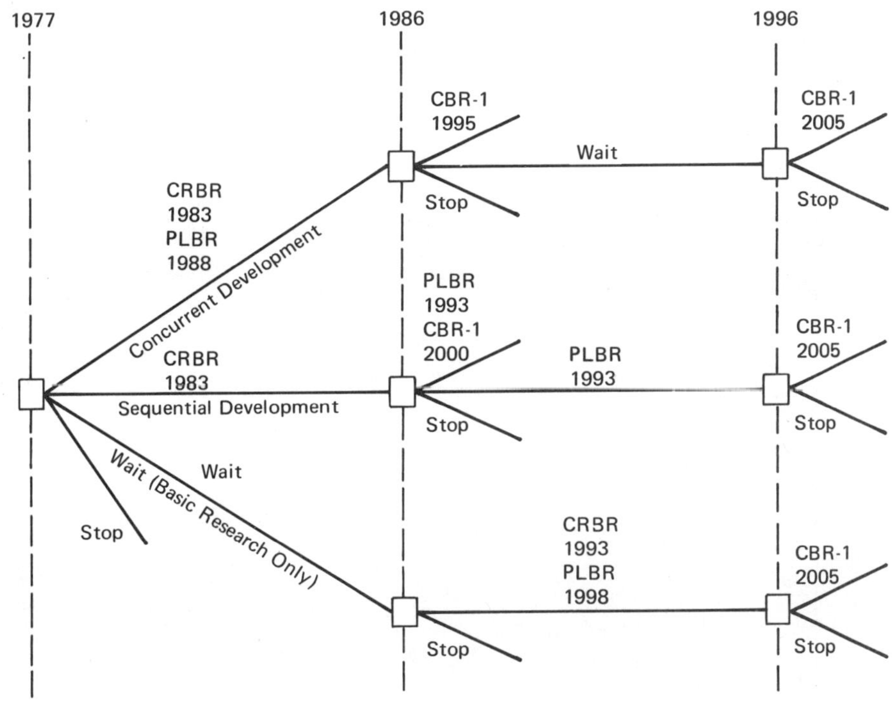

FIGURE IV.2 Alternative breeder R&D strategies.

TABLE IV.5 Dates of Resolution of Key Uncertainties and Judgmental Probability Estimates (Numbers in Parentheses Are the Judgmental Probability Estimates for the Respective Outcomes)

1986		1996		Later	
1. Demand[a]	H (0.29) M (0.31) L[b] (0.4)	5. Uranium resources	low (0.06) intermediate (0.17) high[a] (0.77)	6. 6 GW solar electric in 2000	2.7 × LWR (0.78) 1.25 × LWR[a] (0.22)
2. Coal and shale constraints	50 Quads (0.62) (0.38) none[b]				
3. Cost of nonelectric AES	\$8 (0.27) \$5 (0.55) \$2[b] (0.18)				
4. Public acceptance of breeder	yes (0.54) no[b] (0.46)				

[a]For an explanation of how these probabilities were constructed, see footnote *e* at end of Table IV.1.
[b]Zero breeder benefit.

in 1995, and thereby permits the construction of 6 GW of demonstration plants in the year 2000. This path would require a present decision to build the Clinch River Breeder Reactor (CRBR) and also the Prototype Large Breeder Reactor (PLBR) concurrently, and then in 1986 a second decision giving the go-ahead to CBR-1. With this time sequence, the construction of PLBR would begin before CRBR reached criticality, and construction of CBR-1 would begin before the PLBR is completed. Hence, the upper path is labeled concurrent development. Note that this path also includes the option in 1986 of delaying CBR-1 until 2005. In this case, 6 GW of LMFBR capacity would not be available until 2010.

Alternatively, the middle path (sequential development of CRBR and PLBR) can be followed. Here, the demonstration plant is completed 3 years before a decision is required on the commercial prototype. Then in 1986, we have three options: concurrent development of PLBR and CBR-1, continued sequential development, and stop. The first permits CBR-1 availability in the year 2000, whereas continued sequential development delays CBR-1 availability until 2005.

The third strategy involves waiting until 1986 before making a decision on CRBR and PLBR. Then, if the available information favors proceeding with the program, the concurrent development route is followed. The fourth strategy consists of stopping now. By definition, this yields net economic benefits of zero.

Resolution of Key Uncertainties

The next step in the decision analysis process is to incorporate the major uncertainties into the decision flow diagram. Table IV.5 lists the key uncertainties along with possible outcomes. Some of the alternative outcomes indicated in the survey were deleted in order to simplify the analysis. Note that there are two key future dates for decisions to be made. These future decisions can be based upon the improved information that will be available one and two decades hence (that is, in 1986 and 1996).

The breeder R&D problem is summarized in Figure IV.3. Notice that four sets of *chance* nodes (denoted by circles) have been added to the original decision-flow diagram. The uncertainties resolved at each chance node are listed at the bottom of the diagram.

As of 1977, we must choose one of four program options. Currently, we are uncertain about each of the key parameters. By 1986, however, it is assumed that four areas of uncertainty will be resolved: energy demand growth, coal and shale constraints, the cost of a nonelectric AES, and the public acceptance of the breeder technology. Since the resolution of these uncertainties is independent of the 1977 decision, identical chance nodes appear on each path. The symbol  is shorthand for a transition to one of the thirty-six possible outcomes.

In 1986 two areas of uncertainty remain: uranium supply and solar-electric capital costs. Chance nodes succeeding 1986 decision nodes are no longer independent of the previous path. If a "CBR-1 1995" or "CBR-1 2000" path is taken in 1986, no further decision nodes remain. Consequently, all the remaining uncertainties can be represented in a

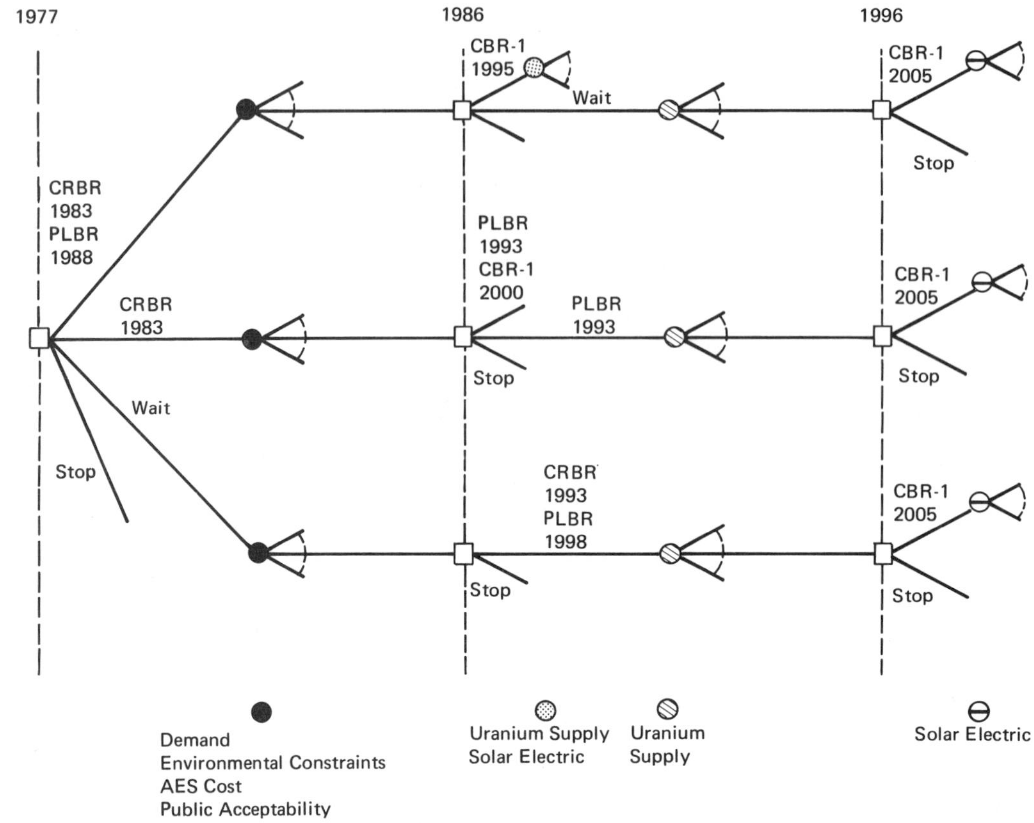

FIGURE IV.3 Decision-flow diagram for breeder R&D.

single chance node, the future outcome of which will affect ultimate breeder benefits but will not affect decisions made regarding R&D for these branches. The shorthand symbol denotes (see Figures IV.3, IV.4, and IV.5):

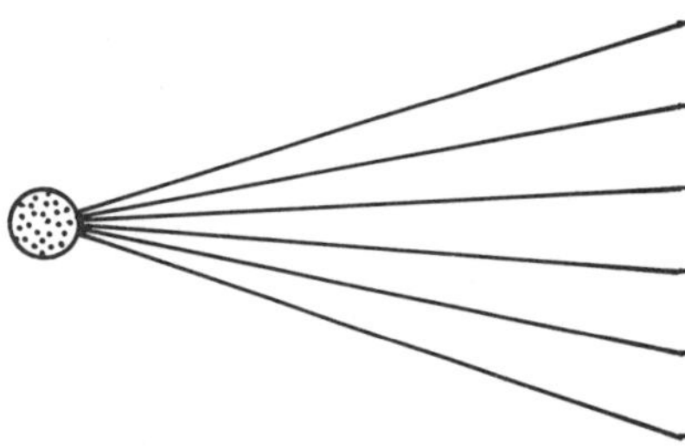

	Uranium Supply	*Solar Electric*
	high	1.25 × LWR
	high	2.7 × LWR
	intermediate	1.25 × LWR
	intermediate	2.7 × LWR
	low	1.25 × LWR
	low	2.7 × LWR

But if a 1986 decision leads to a 1996 decision node, we can still take advantage of the sequential resolution of uncertainty. In this case, the uncertainty regarding uranium supply will be resolved before the final decision date. The symbol denotes:

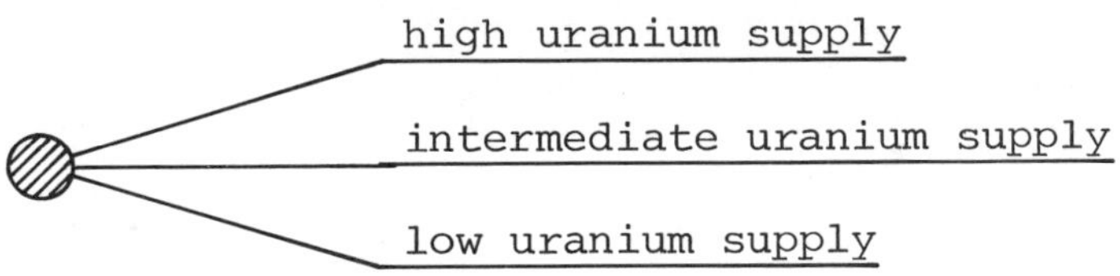

The remaining uncertainty is then resolved in the final chance node after the 1996 decisions have been taken. The symbol  denotes:

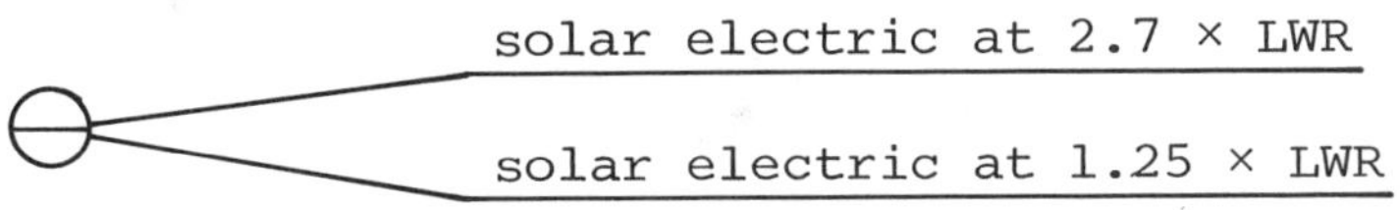

Probabilities for Alternative States-of-the-World

The judgmental probability estimates summarized previously were used to calculate the probabilities for states-of-the-world at each chance node. These are summarized in Table IV.5 with modifications made to certain parameters listed in Table IV.1 for purposes of simplification. It will be noted that the possible outcomes for public acceptance of nuclear option and for the costs of competing technologies such as solar and fusion have been simplified. For these first calculations, it was assumed that the key uncertainties are uncorrelated. Hence, the probability of a scenario is simply the product of the probabilities of the individual events. Using the mean survey responses of the nine MRG members and consultants, the probability of zero breeder benefits is 0.97. This probability assumption is the key to understanding the numerical results that will be reported below. *Other probability assessments could lead to very different conclusions on the economic benefits of the breeder program.*

Expected Net Economic Benefits

The next step is to calculate expected net benefits for each alternative 1977 decision. This is done through the process of "backwards induction." Figure IV.4 illustrates the process for the concurrent development option. (Note that the truncated diagram provides details on only one of the many possible sequences of decisions and events.)

The procedure to calculate the expected value of a particular strategy is to start at the end of the decision tree and work back towards the beginning. In the case of Figure IV.4, we start by imagining that we are actually at the 1996 decision node, that is, it is assumed that the concurrent development option was selected in 1977 and then in 1986 a decision was made to wait for better information regarding uranium supply before making a decision on commercialization. It is further assumed that uranium supply turned out to be low, energy demands high, a 50-Quad constraint on coal and shale, an $8 AES, and public acceptance of the breeder technology. Then in 1996 there would be two alternatives:

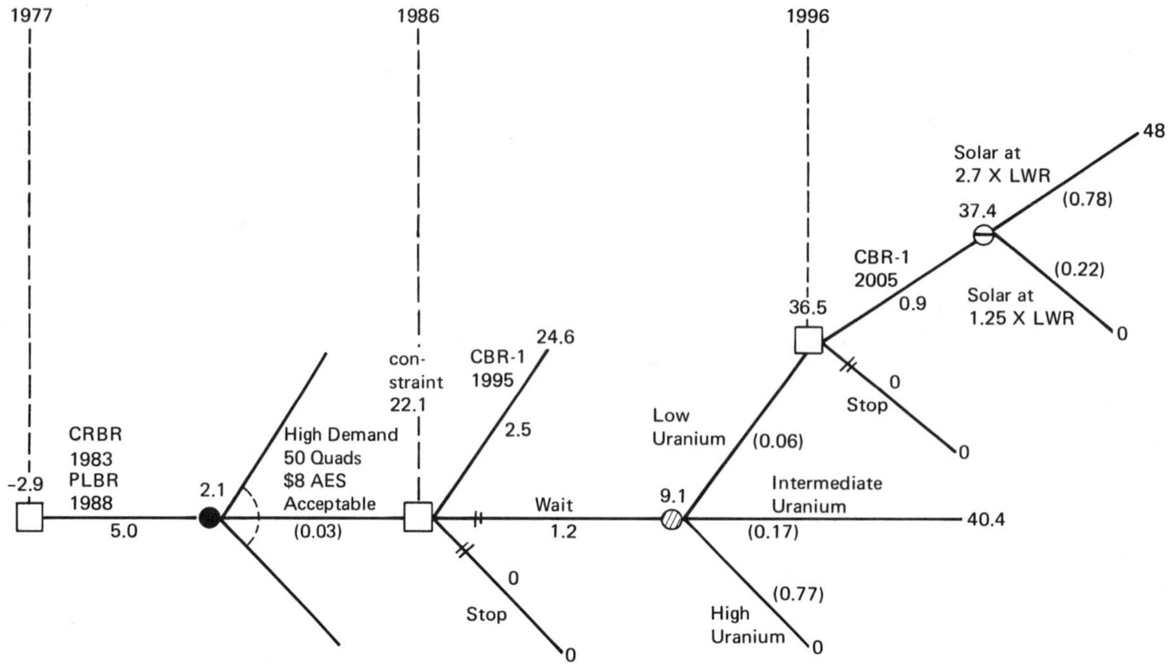

FIGURE IV.4 Truncated decision-flow diagram for concurrent development ($ billion discounted to 1975 at 6 percent per year).

commercialization or termination of the development program. To explain the procedure for making this decision, we isolate that part of the diagram from the 1996 decision node onward as shown in Figure IV.5. Note that the probabilities of outcomes are placed in parentheses on the branches emanating from the chance node, and the gross benefits for each sequence of decisions and chance events are presented at the tips of the decision tree. The underscored numbers on the branches emanating from the decision node represent the additional R&D costs resulting from a particular decision. See Appendix F.2 for a discussion of R&D costs over time that were assumed in this study.

The expected value of the decision to commercialize in 2005 may now be determined. Gross benefits from commercialization are conditional upon solar-electric capital costs. If solar electric is available at 2.7 times the capital costs of the LWR, then breeder benefits will be $48 billion (see Table IV.2). However, if solar electric is available at the lower capital cost (1.25 times LWR) breeder benefits will be zero. Given our set of nominal probabilities, expected gross benefits are

$$0.78(\$48) + 0.22(0) = \$37.4 \text{ billion}$$

Subtracting $0.9 billion for the added R&D costs, expected *net* economic benefits from commercialization are $36.5 billion. If the program is stopped in 1986, however, no additional R&D costs are incurred and gross benefits are zero. Hence, within the domain of economic considerations, the commercialization strategy is clearly preferable. The double slashes

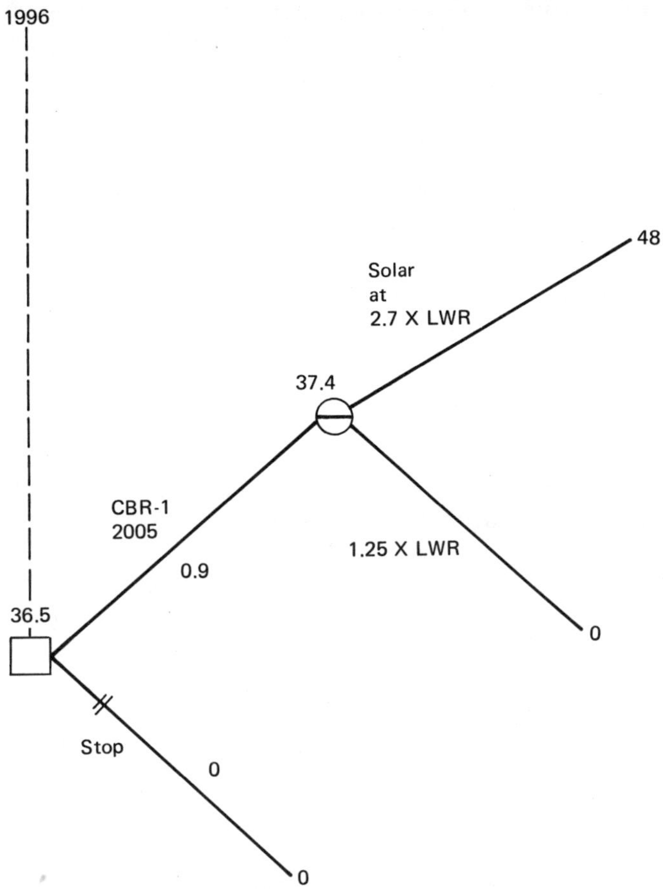

FIGURE IV.5 Decision analysis of 1996 decision ($ billions discounted to 1975 at 6 percent per year).

on the stop path indicate that the path is closed. The expected value of the preferred decision is placed above the decision node in Figure IV.5. Thus, by computing the expected value at the chance node and folding back to the previous decision node, we are able to calculate the optimal 1996 decision under uncertainty.

Next we step back to the 1986 "wait" decision. The expected gross benefits from this decision are $36.5 billion given low uranium supply. Similarly, expected gross benefits are $40.4 billion, given intermediate uranium supply, and zero with high uranium supply. Hence, expected gross benefits from a 1986 "wait" decision are

$$0.06(\$36.5B) + 0.17(\$40.4B) + 0.77(0) = \$9.1B$$

and expected net benefits are

$$\$9.1 - \$1.2B = \$7.9B$$

Continuing the process of backwards induction, in 1986 we are faced with three choices: commercialization in 1995, wait for better information, or stop. Given high demand, a 50-Quad limit on coal and shale, an $8 AES, and public acceptability of the breeder technology, the economic considerations tend toward 1995 commercialization as the best choice.

Finally, averaging out and folding back to 1977, we find that the concurrent development option has negative expected net benefits.

Applying this process to the other alternatives, the initial decision reduces to the one in Figure IV.6. Thus, given the MRG set of judgmental probabilities set forth in Table IV.5, stop is the economically optimal strategy. The expected benefits have a probability distribution for each decision option that is related to the uncertainties. In this case, the standard deviation of that distribution is about $20 billion.

As the expected benefits and R&D costs are combined with probabilities at each node in the decision tree, those costs are expressed in economic terms. It is recognized that there are other social costs or externalities that might be considered. Among these are biomedical and environmental effects, balance of payments in trade, and national and international security. At the present state of the art, these costs are not expressed in economic terms and therefore have not been considered in the backward calculation of costs and benefits. These considerations do in part show up in the analysis as additional uncertainties, e.g., public acceptability of the breeder, and level of oil imports. This treatment assumed independence of the probabilities attached to the various uncertainties. Further work is recommended to estimate the degree of dependence among the various uncertainties and to determine the effect of this simplifying assumption. (See p. 133, item 5.)

An Alternative Analysis Based on a Revised Uranium Resource Assessment

The preceding demonstration analysis has been based upon the personal (judgmental) probabilities of the nine second-round respondents to the poll conducted within the MRG and consultants to that group. Soon after

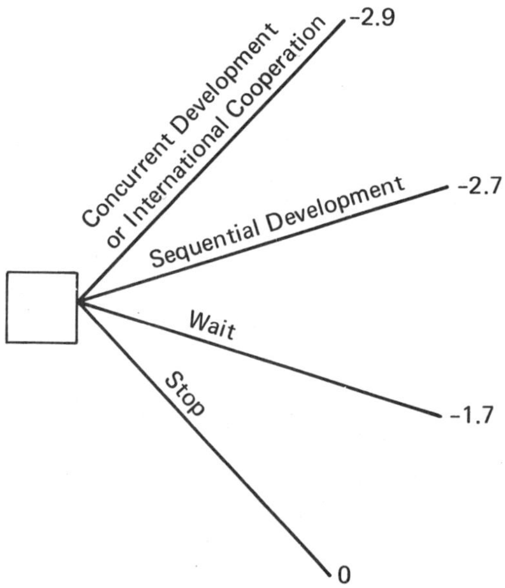

FIGURE IV.6 Expected net economic benefits of alternatives ($ billions discounted to 1975 at 6 percent per year).

the MRG's initial results were circulated, it became evident that the CONAES Uranium Resources Group (URG) held a considerably more pessimistic view with respect to uranium availability.

Decision analysis is an iterative process, and improved information can readily be factored into successive calculations. In a complete policy study using this methodology, specialists should be consulted concerning each individual area of uncertainty. Although this was not possible in this study for each critical parameter, two members of the MRG did have an opportunity to meet with the URG, and to elicit that group's estimates of uranium availability. This parameter was known to be an important one, and it is instructive to outline the interview procedure as an example to be followed or improved upon in future applications. The results were incorporated into a revised analysis that resulted in positive net benefits for the concurrent development strategy. These results are supplemented by sensitivity analyses that illustrate the robustness of the optimal decision to one-at-a-time variations in the judgmental probabilities attached to the remaining variables.

The interview procedure involved three phases, an outline of the decision analysis approach and objectives followed by two rounds of estimates. The explanation of the methodology and objective was found to be quite important, and considerable time was devoted to developing an unambiguous phrasing of the definition of the question. A cumulative probability estimate was ultimately employed and is clearly superior to the phrasing of the question in terms of the point probability estimates used in the MRG survey. In any case, point estimates are used in the analysis of the outcome.

As a result of the discussions, there emerged a probabilistic assessment that may be described as a "consensus" view among the five URG participants who met in Tucson, Arizona, on December 1, 1976. The URG's views may be summarized as follows (Cazalet-Hogan memorandum of that meeting, dated December 2):

> In developing these estimates, the group recognized that they are conservative relative to some other estimates, but they felt that these were best available estimates given the lack of data and the extreme extrapolations required for this assessment. These estimates are, however, in the nature of highly subjective judgments drawn from the panel members over a period of a few hours and without time for reflection and consultation with other group members not present.
>
> Of particular importance to this assessment was the view that there may be little uranium ore in the concentrations of sandstone deposits which might bear on the $100 per pound (forward cost) deposits because of physical and chemical reasons. If this were true, an increase in price would have relatively small effect on total resources. The estimates shown here reflect this view. The probability of less than 1.8 million tons is only slightly higher in the $30 case as compared to the $100 case for each of the members. On the other hand,

at the high end of the probability distribution, resources increase from roughly 4 million tons at $30 per pound to 10 million tons at $100 per pound. In other words, the group assigns a small, but significant, probability to large increases in price bringing about a large increase in resources over a long period of time.

The URG's "consensus" view on total resources available at a forward cost of less than $100 through the year 2035 was described through the following cumulative probability distribution:

X million short tons:	1.8	3.0	5.0	10.0
Probability that uranium resources are less than X:	0.2	0.7	0.95	0.99+

Note that the URG's assessments were represented in the form of a continuous probability distribution. The decision analysis procedure employed here requires that *discrete* probabilities must be assigned to the three representative uranium supply curves shown in Table IV.2. At a selling price of $150 per pound (forward cost of $100), these three curves imply resource levels of: (a) 2.25, (b) 3.7, and (c) 9.2 million short tons. The first step was to employ graphical interpolation of the MRG consensus estimates to estimate the *cumulative* probabilities associated with these three levels of uranium resources, and then to define discrete probabilities for the three representative points as follows:

X million short tons available at a selling price of $150 per pound (forward cost of $100):	2.25	3.7	9.2
Probability that uranium resources are less than X (through graphical interpolation):	0.35	0.81	0.99
Discrete probability that uranium resources are *equal* to the representative level X:	P_a	P_b	$P_c = 1 - P_a - P_b$

From the continuous probability distribution, the following bounds can be constructed:

$$0.35 < P_a < 0.81$$

$$0.81 < P_a + P_b < 0.99$$

The final step is to assume that these discrete probabilities lie halfway between their upper and their lower bounds. Therefore:

$$P_a = 0.5\ (0.35 + 0.81) = 0.58$$

$$P_a + P_b = 0.5\ (0.81 + 0.99) = 0.90$$

$$\therefore P_b = 0.32$$

$$P_c = 0.10$$

For a comparison of the discrete probability assessments made by the URG and the MRG, we then have:

X million short tons available at a selling price of $150 per pound (forward cost of $100):		2.25	3.7	9.2
Probability that uranium resources are equal to *X*	URG:	0.58	0.32	0.10
	MRG:	0.06	0.17	0.77

Note the significant differences between the URG and MRG estimates. The URG asserts that the more knowledgeable the geologist, the greater the tendency for the lower figure.

In the following decision analysis, the URG's uranium resource probabilities have been employed instead of the MRG's. All other probabilities, however, are identical to those estimated previously by the nine MRG respondents. Hereafter, we shall refer to this revised set of probabilities as the "alternative" set.

Figure IV.7 presents the results of the revised decision analysis. With this alternative set of probabilities, concurrent development is the optimal strategy. Expected net benefits are now approximately $4.5 billion.*

It was not possible to obtain probability assessments from outside consultants concerning the remaining uncertainties. However, we can calculate how the optimal strategy depends upon these other uncertainties. For this purpose, we have varied the judgmental probabilities one at a time--assigning the alternative probability values to all other random variables. This provides a measure of the robustness of the optimal initial decision to other probability assessments.

Figure IV.8 shows the effect of varying the probability of low-energy-demand growth.† (Low-energy-demand growth is one of the possible

*Because of time and computer budget limitations, we have had to make a few numerical approximations in order to compute these results by hand rather than through a formal decision tree program.

†The high and intermediate growth-rate probabilities are varied proportionately to each other as the low-demand probability is changed. A similar procedure was followed in each case where more than two outcomes are logically possible.

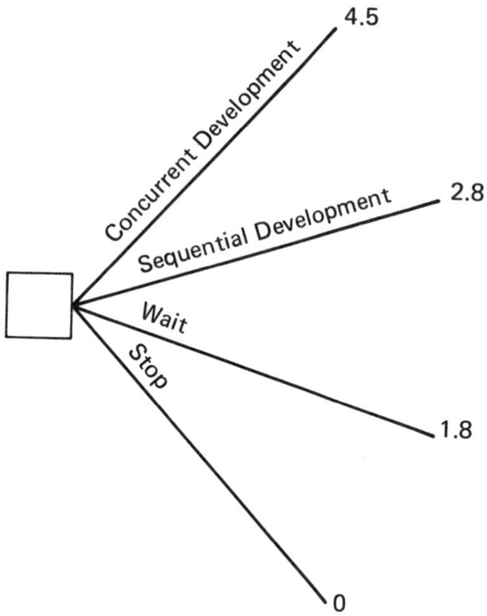

FIGURE IV.7 Expected net economic benefits. Revision--based on URG uranium probability assessments ($ billions discounted to 1975 at 6 percent per year).

events that could lead to zero gross economic benefits.) According to this calculation, expected net benefits from concurrent development rise to $10.8 billion when the probability of low demand growth falls to zero. When selected outcomes are assigned a probability of zero, the remaining probability was allocated among other outcomes of that uncertainty in proportion to the original estimates. Alternatively, the losses from concurrent development would be $5 billion (the R&D costs from 1977 to 1986) if the probability of low demand were 1.

Table IV.6 summarizes the results of a similar sensitivity analysis on each of the remaining random variables. For reference purposes, this table also includes the values used for each of the critical uncertainties, together with the overall probability of one or more events that could lead to zero gross economic benefits. Overall, this probability is 0.88, according to the MRG-URG alternative assessment.

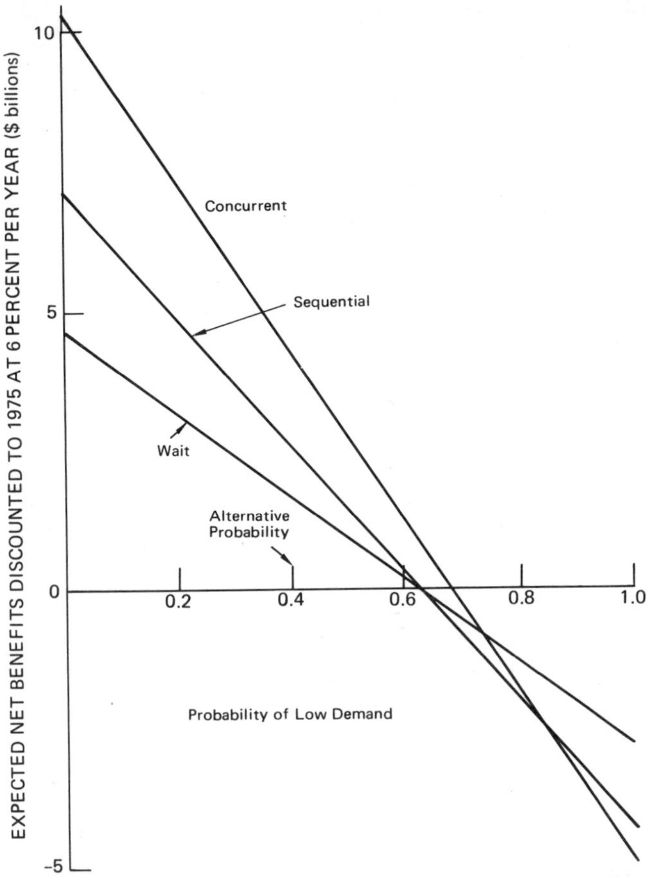

FIGURE IV.8 Effect of low-demand probability on economically optimal decision.

IV.5 ANALYSIS OF ADVANCED NUCLEAR REACTOR RESEARCH AND DEVELOPMENT USING THE NORDHAUS MODEL

This section presents a comparison of net economic benefits of alternative reserach, development, and commercialization strategies (abbreviated as R&D) for advanced nuclear systems. The emphasis is on the timing of availability of two types of advanced nuclear reactors.

The decision structure and sequential nature of the decision process were simplified somewhat in comparison with the analysis of Section IV.4. As far as the decision structure is concerned, the question asked here is: First, what are the economic benefits of alternative decisions net of R&D costs but before accounting for environmental costs? In what follows, this accounting concept is called the "net economic benefit." The decisions analyzed are whether either, both, or neither of the LMFBR and an intermediate technology, the ACR, should be pursued if the net economic benefit is the only criterion applied. The two possible reactors will be called *advanced nuclear* in the present section. For each reactor, the options analyzed are R&D efforts pursued in such a way that they will be completed in one of five periods--2000, 2010, 2020, 2030, and "never." Thus, the problem is to calculate the net benefits for each of 25 possible decisions.

TABLE IV.6 Probabilistic Sensitivity Analysis Effect of One-at-a-Time Variation of Critical Variables

Realizations (or Policies), Any One of Which is Judged to Lead to Zero *Gross Economic Benefits*	MRG[a] Probability of Realization	Expected Net Benefits of Concurrent Development If: Probability of Event[a] = 0 ($ billion)	Probability of Event[b] = 1 ($ billion)
Low demand	0.40	10.8	-5.0
No coal and shale constraints	0.38	10.3	-5.0
$2 AES	0.18	6.6	-5.0
Breeder unacceptable	0.46	12.6	-5.0
High uranium resources	0.10	5.6	-7.5
Solar electric at 1.25 × LWR cost	0.22	7.2	-7.5
Probability of one or more realizations leading to zero *gross* economic benefits	0.88		
Expected *net* benefits of concurrent development	$4.5 billion		

[a]Calculated from responses to questionnaire given in Appendix F.1.
[b]Given alternative values of all other probabilities.

This analysis is carried out with the Nordhaus model. The attention is primarily on the "downstream" decisions. That is, what are the benefits of having R&D completed by one of the dates mentioned above? As in all the MRG work, the R&D comprises all research, prototype or demonstration plants, and early commercial plants up to having 6 GW^e installed capacity. Thus the analysis is relatively aggregated and does not investigate the important detailed planning problems involved in reaching the 6-GW^e level. This approach can be rationalized as follows: Before a detailed examination of exactly how a given option should be implemented (where should plants be built, who will be the vendors, what form of pricing or taxation should be used to pay for the R&D, etc.?), an aggregate assessment of each option like that of the MRG work should be made. If the aggregative assessment is definite one way or the other, the results will give a good indication as to what detailed implementation plans are likely to be fruitful.

The analysis dealt with the same eight uncertainties that were studied in the survey of MRG members and consultants described in Section IV.2. In what follows, we will refer to each particular resolution of the uncertainties as a *state-of-the-world* (also called "realization" in Chapters II and III); thus the MRG "Base Case" is one of a total of 10,368 states-of-the-world that could be examined. It should be stressed that, by explicitly treating these eight uncertainties, we have opened up the analysis to an enormous number of possible realizations (states-of-the-world), and also that there are other uncertainties that are not treated and will affect the outcome (such as the possible breakup of OPEC, breakthroughs in abatement techniques for generating electricity from coal, or a large-scale war). In some cases we have explicitly tested for importance of variables; in others, omission is due either to prior expectation or simple lack of time and imagination.

In summary, the purpose of the R&D analysis conducted here is to respond to the question: What is the net economic benefit of each of the possible decisions about advanced nuclear technologies in each state-of-the-world? Also, given a set of judgmental probabilities, what is the expected value of the benefit of each of the decisions?

Description of the Technique for Estimating Benefits

We can only sketch the detailed steps involved in making the calculations, with further details available in the backup material.* The process of estimating the value of decisions about advanced nuclear reactors involved four steps.

First, among the 25 × 10,368 different outcomes (25 decisions and 10,368 states-of-the-world) a small number of runs were chosen, giving eventually approximately 100 "observations." The runs were chosen so as to minimize the expected error in the end results of our calculations. This meant sampling more heavily those runs where the expected

*"The Value of Advanced Nuclear Power," W. D. Nordhaus, duplicated January 13, 1977.

value of advanced nuclear was high and those runs which were relatively likely. In some cases the choice was judgmental, while other runs were clearly seen to be important from past runs, or from the preliminary probability questionnaires. The chosen runs were then made using a slightly modified version of the earlier Nordhaus model. The main modification was the introduction of an integer-programming code, so that in each state-of-the-world the calculation would always find the decision that yielded the highest net benefit. From these runs were obtained the estimated value of R&D decisions conditional on the state-of-the-world.

Second, it was necessary to interpolate between different runs for the missing observations. This step proceeded in two stages: first, from *a priori* information about the structure of the underlying economic problem, a small model was constructed which gave the "predicted value" of a policy decision conditional on the state-of-the-world. In order to preclude any possibility of *a priori* ideas entering into the procedure, this "small model" was not based on any runs. It essentially estimates the demand for advanced nuclear technology (as a function of GNP growth and of the presence or absence of coal and shale limits) and the supply function of AES's (coal, LWR, solar, or fusion). The value of an advanced reactors decision was then estimated by the product of the demand for advanced nuclear technology times the cost advantage (if any) of that technology over the AES.

Because the "small model" had much smaller dimensionality than the large model, the "predicted values" had correspondingly larger errors. In a second stage, a regression analysis was performed in which the "actual values" (i.e., the values coming from the selected model runs) were regressed on the values predicted by the small model as well as on decision and realization variables. This procedure tightened up the fit considerably, so that in a weighted logarithmic regression taken within the sample of selected model runs, the standard error of estimate was approximately 10 percent of the actual value. In making the final calculation of the third step, discussed below, the predicted values from the regression model were used as estimates of the value of each decision in each state-of-the-world. It is important to note that the regression technique ensures that the weighted sum of the prediction errors in the regression is exactly zero, so no systematic errors are introduced by this procedure.

The third step of the analysis combines the information from the mathematical programming and regression analyses discussed in steps one and two above with judgmental probability estimates to calculate the expected value of the decisions. The procedure described in step two allows us to estimate the value of all decisions for all the different states-of-the-world. We use the results of the judgmental probability questionnaires* to provide estimates of the probability of each state-of-the-world. Combining these, we obtain the distribution of the net benefits of each of the decisions net of R&D cost but before considering environmental costs.

*Reported from the second round.

A final detail in the procedure is the question of sequential revelation of the outcomes of uncertain events. Each of the eight uncertainties shown above will be resolved at some point, and once that uncertainty has been resolved, the expected value of decisions will generally change and the rank ordering of decisions may also change. At present we have treated them in a very simple manner: it is assumed that *none* of the uncertainties will be resolved before the decision about advanced nuclear systems is made and fully implemented. For example, if we wish to decide about capital cost of the LMFBR, it is assumed that we must proceed through the R&D on the LMFBR before we can "buy" that knowledge. Similarly, it is assumed that the acceptability of the advanced converters will not be known before the full R&D decision has been made.

It is clear that this assumption about resolution of uncertainty is extreme; in particular, it is probable that a great deal will be learned about uranium supplies, demand growth, and coal technologies in the next 20 years. At a further point in this chapter, we test the difference that earlier resolution of uncertainties makes. We do, however, have some general idea of the importance of resolution of uncertainty. For example, the uncertainty about energy demand functions, as well as about coal and other nonelectric technologies, will probably be resolved independently of the decision about advanced nuclear power, and these variables are thus exogenous uncertainties. In these cases, then, there is value in postponing decisions about advanced reactors, for the "insurance premia" will decline. Other uncertainties--such as the two mentioned in the preceding paragraph--are probably integrally linked to the outcome of the R&D effort itself. These R&D uncertainties will *not* completely evaporate over time, and there is less value to postponing decisions than in the case of the exogenous uncertainties.

In general then, the assumption of no revelation of uncertainties is likely to bias the estimates in favor of early decisions. Without further analysis it is not possible to say how large the bias might be. A clear effect of this assumption is to ignore the possibility of terminating the R&D and/or commercialization program if it should become evident at some future date that the breeder is not competitive. Thus in this regard the effect is that of reducing the expected net benefits.

Results for the R&D Analysis

A preliminary set of results for the R&D analysis is presented in this section. Moreover, it may be useful to add further "observations" or to change some of the parameters of the "small model" in order to improve the accuracy of the approximation. Finally, the probabilities are bound to change in future rounds of collecting estimates.

Table IV.7 gives the different states-of-the-world, and the probabilities associated with them. These are drawn from the results of the judgmental probability estimates survey of Section IV.2. Note that in several cases, the probabilities had to be interpolated, extrapolated, or interpreted in order to conform to the specifications of the Nordhaus model.

TABLE IV.7 States-of-the-World (SOW) Investigated in R&D Analysis

I Index	Values and Mean Probabilities (Section IV.2) SOW(I,1)	SOW(I,2)	SOW(I,3)	SOW(I,4)
1 *Coal/shale limits*				
Production limit (Quad/yr)	∞	62.0		
Probability	0.38	0.62		
2 *GNP growth*				
Average growth rate 1975-2030, percent per annum	1.2	3.1	4.1	
Probability	0.26	0.57	0.16	
3 *Nonelectric AES*				
Cost, $ per million	$2	$5	$8	
BTU Probability	0.18	0.56	0.27	
4 *Electric AES* (best of solar or fusion)				
Capital cost as multiple of LWR	1.25	1.6	2.7	4.0
Probability	0.44	0.25	0.23	0.08
5 *Uranium resources*				
10^6 tons up to $150/lb	2.1	5.0	9.2	13.5
Probability	0.06	0.17	0.40	0.37
6 *Capital cost, AC*				
Cost as multiple of LWR	1.0	1.1	1.25	
Probability	0.15	0.52	0.33	
7 *Capital cost, LMFBR*				
Cost as multiple of LWR	1.0	1.25	1.5	
Probability	0.10	0.56	0.34	
8 *Public acceptance of advanced nuclear*				
Event	No advanced nuclear	AC only	LMFBR only	ALL
Probability	0.30	0.16	0.14	0.40

SOURCE: MRG probability assessment questionnaire, second round.

Table IV.8 shows the undiscounted and discounted costs assumed for R&D on the different reactors. The costs for one reactor are assumed to be independent of the decision for another. The cost figures are subtracted from the gross benefits of each technology to obtain the net benefits. Note that the net benefits with alternative costs of R&D may be easily calculated below by simply adding or subtracting the difference between any assumed value and that in Table IV.9 below.

Finally, Table IV.9 summarizes the net benefit estimates for each of the 25 combinations of decisions. Along the top of the table are the decisions about the date of introduction of the LMFBR. In the center are results for the *discounted net economic benefit* of different decisions. As in all MRG work, all costs and benefits are discounted at a rate of 6 percent per annum. For each entry is given the expected (or mean) for each of the decisions. Although there is no exact way of calculating the "standard error" of the estimates, from the regression analysis, the judgmental estimate is that the standard error of estimate is approximately $1 billion.

The overall picture that emerges from Table IV.9 is that, given the assumptions of the present analysis, there appears to be good reason to postpone implementation of R&D on either of the two advanced reactors.

TABLE IV.8 Costs of Decisions to Implement Different Reactor Types, Billions of Dollars, 1975 Prices

	Advanced Converter	LMFBR
Undiscounted (1975 prices)		
R&D	5	10
Commercialization	10	10
Costs, discounted (at 13 percent per annum) to 1975 (1975 prices) for 6 GW^e in place by:		
2000	5.9	8.7
2010	3.3	4.9
2020	1.8	2.7
2030	1.0	1.5
2040 or later	0.0	0.0

SOURCE: K. Hoffman, D. Cope, and R. Richels for the undiscounted value. The discounted values make the following assumptions:

(a) Commercialization occurs during the period from 10 to 0 years before the availability date. The cash flows are assumed all to be concentrated 5 years before the availability.

(b) Research and development is assumed to occur from 20 to 10 years before the availability date. The cash flows are assumed to be concentrated 15 years before the availability date.

TABLE IV.9 Net Economic Benefits of Decisions About Advanced Nuclear Reactors, Exclusive of Environmental Costs but Subtracting R&D Costs, Discounted to 1975, at 6 Percent Per Annum, in Billions of Dollars at 1975 Prices

Expected Values of Decisions

LMFBR Decision: Available at 6 GWe Capacity by	Advanced Converter Decision: Available at 6 GWe Capacity by 2000	2010	2020	2030	2040 or Later ("Never")
2000	-7.4	-6.4	-5.3	-4.3	-2.6
2010	-3.6	-2.6	-1.6	-0.5	1.1
2020	-1.4	-0.5	0.5	1.6	3.0
2030	-0.4	0.5	1.5	2.4	3.6
2040 or Later ("Never")	0.3	1.2	2.1	2.9	0.0[a]

[a]Note that the never-never decision is the reference path; therefore, it has zero value in all cases.

The best case of immediate development--where "immediate" means having advanced nuclear technology on line by 2000--is to develop the advanced converter but not the LMFBR. According to Table IV.10 the net economic benefits of the advanced converter in 2000 are $0.3 billion, while those of the LMFBR are -$2.6 billion.* A crash program to develop both advanced reactors is the least economic of the decisions, having a net benefit of -$7.4 billion. The economically best program that includes both reactors is to have both ready in the year 2030, yielding $2.4 billion.

Using only the criterion of net economic benefits, the highest expected value comes from a decision that implements the LMFBR in 2030, while the advanced converter is not developed. This strategy has an expected net benefit of $3.6 billion. Other strategies in the neighborhood of the optimal one, however, are hardly less economic: the decision to speed up the LMFBR 10 years from the point selected as optimal by our criterion loses only $0.6 billion, while the decision to develop the advanced converter rather than the breeder in 2030 loses $0.7 billion.

It is interesting to note that the current U.S. strategy--to develop the LMFBR by 2000 but not to develop the advanced converter--ranks nineteenth out of the 25 decisions that were considered.

*The normalization used here, it must be emphasized, is that the value of a decision not to have any advanced nuclear technology is arbitrarily set at zero. Therefore all net benefit calculations are the difference between the benefits from the decision under consideration and the benefits of no advanced nuclear technology.

TABLE IV.10 Net Economic Benefits of Decisions About Advanced Nuclear Reactors, Exclusive of Environmental Costs but Subtracting R&D Costs, Discounted to 1975, at 6 Percent Per Annum, in Billions of Dollars at 1975 Prices

Standard Deviation of Outcome

LMFBR Decision: Available at 6 GW^e Capacity by	Advanced Converter Decision: Available at 6 GW^e Capacity by				
	2000	2010	2020	2030	2040 or Later ("Never")
2000	20.6	15.0	13.5	15.0	20.7
2010	20.9	17.3	13.6	15.0	20.3
2020	21.0	17.3	13.4	14.6	19.3
2030	20.9	14.9	13.0	13.8	17.0
2040 or Later ("Never")	20.4	14.4	12.2	12.1	0.0[a]

[a]Since the reference point is always the path without any advanced nuclear, the never-never decision has a zero standard deviation. All other paths thus measure the variability of the outcome relative to the never-never decision.

A second feature of the analysis that is worth attention is the degree of variability of the outcome. Table IV.9 shows the standard deviation of the outcome for each of the decisions relative to the no-advanced-nuclear path. The standard deviations are quite large--in the order of $20 billion--indicating that the effect of policy on the discounted value of real income has a large dispersion across states-of-the-world.

The dispersion of the outcomes is shown graphically in Figures IV.9 through IV.11. Each of the figures shows the frequency distribution of net benefits according to the probability assessments of the group reported above. Figure IV.9 shows the frequency distribution for the optimal decision (LMFBR in 2030, advanced converter in 2040 or later). As in all the decisions, there is a very sizable fraction of the distribution at the far left; this indicates, in Figure IV.9 for example, that there is a 56 percent judgmental probability that the value of the new technology is zero *before* subtracting the R&D costs. There is a sizable fraction of the cases, however, that have positive benefits. And at the far right of the frequency, at "100 or more," it can be seen that there is a probability of about 1 percent that the net benefits will exceed $100 billion.

Figure IV.10 shows the frequency distribution for the decision to implement both advanced nuclear reactor types by the year 2000. The major difference between this frequency distribution and that for the optimal decision in Figure IV.9 is that the distribution is displaced

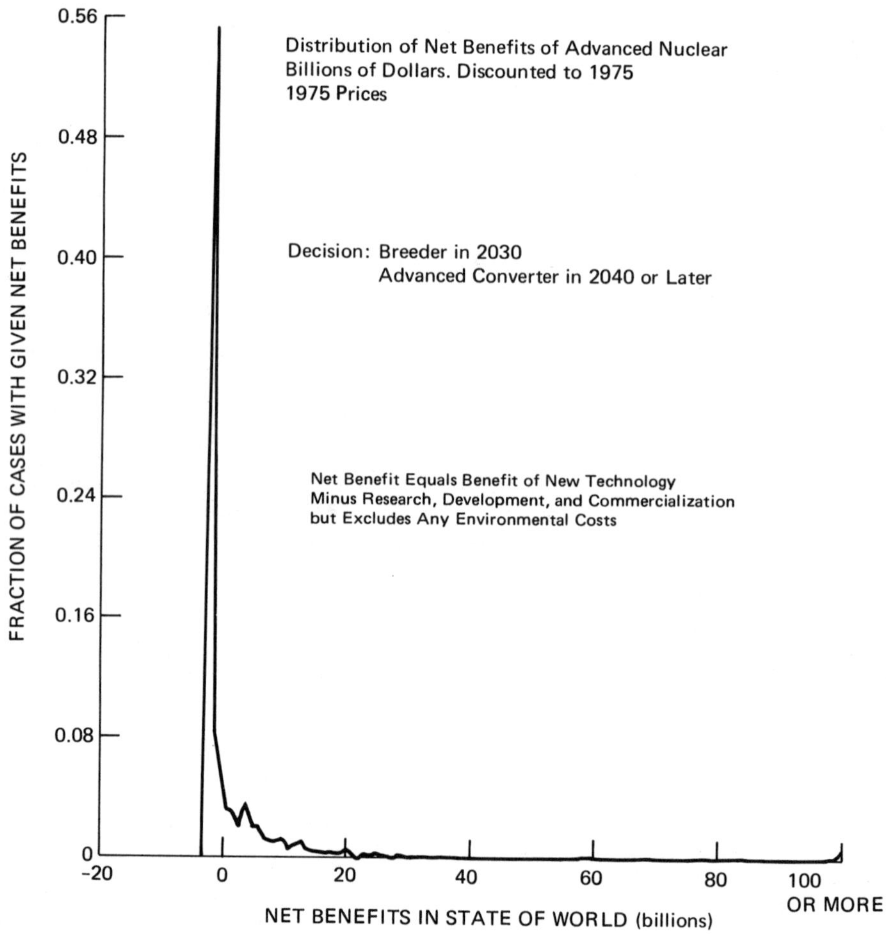

FIGURE IV.9 Distribution of net economic benefits in optimal decision.

to the left (i.e., toward the negative values) by approximately $10 billion.

The Value of Perfect Information

As noted in the earlier discussion the present analysis assumes that the knowledge about the uncertain variables is revealed only *after* the decisions have been made. In certain cases, this is probably an extreme assumption. In the case of demand growth, for example, a great deal will be revealed before the decision to implement the optimal strategy is far along.

In order to test for the importance of the assumption about the sequential resolution of uncertainty, a further run was made in which it was assumed that all realizations were known before the decision was taken. This procedure can be understood as follows: there continue to be uncertainties about the exact state of the world that will hold in the future. Assume, however, that these uncertainties were resolved

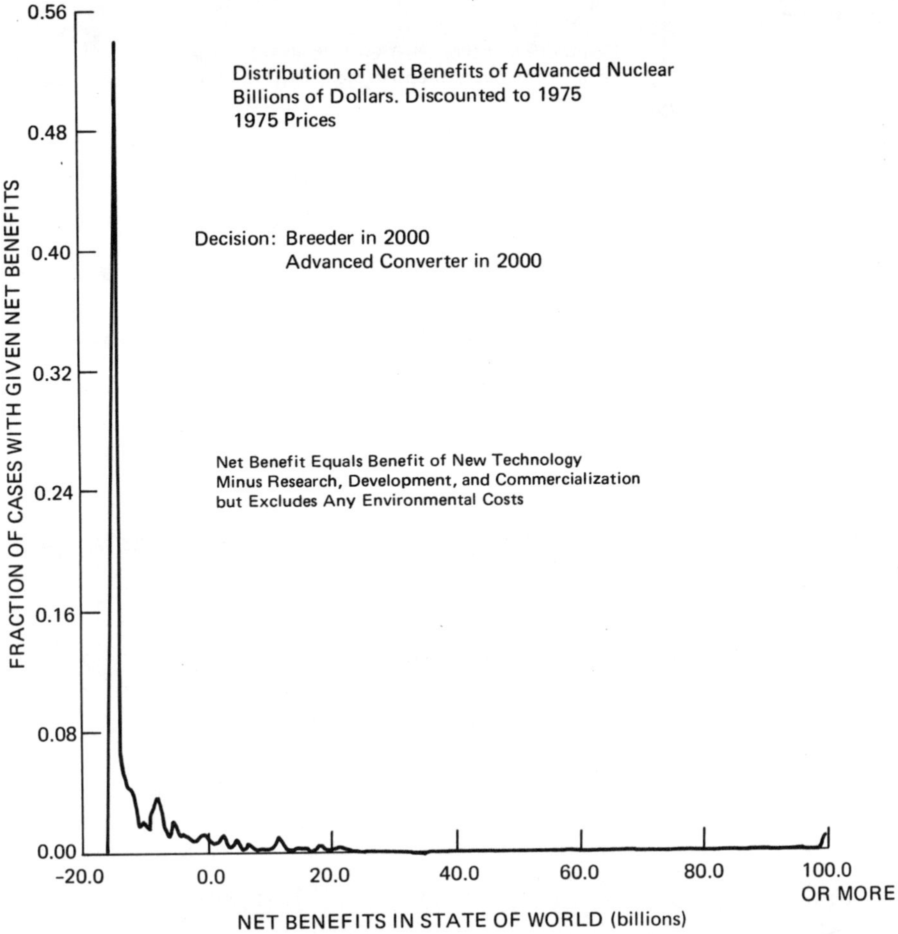

FIGURE IV.10 Distribution of net economic benefits for decision to implement both types immediately.

in the near future, and before any substantial investment in R&D had been made. Then, upon learning which state-of-the-world we are in, decision makers could at that point choose the optimal decision in a deterministic framework.

The question we ask, then, is: What is the economic value of having perfect information about the numerical value of uncertain variables? According to the assumptions used in the present section, the expected value of the decision, when the decision is optimized for each state-of-the-world, is $5.8 billion. Figure IV.11 shows the distribution of net benefits for this case.

How are we to interpret this result? This says that, relative to the optimal decision, the value of perfect information is $2.2 billion. That is, if instead of investing so as to make the LMFBR available in 2030 and never investing in the advanced converter, we had perfect information, this would provide an increase of the benefits from the expected value of $3.6 billion to $5.8 billion, a difference of $2.2

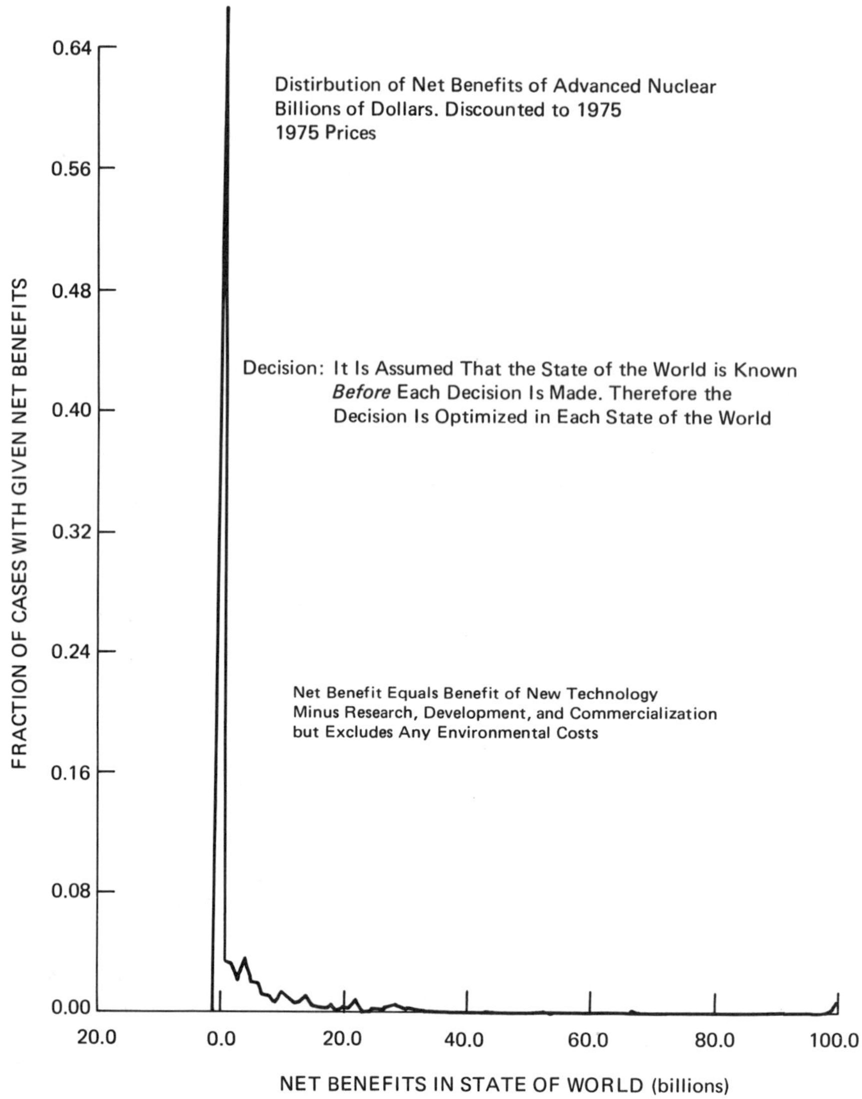

FIGURE IV.11 Distribution of net economic benefits in the optimal decision in the presence of perfect information about uncertain variables.

billion. As can be seen by comparing Figures IV.9 and IV.11, the main value of perfect information would be to avoid investing in R&D for those cases when the net benefits are negative.

Conclusions

The conclusions which can be drawn from the present analysis are as follows: First, the *expected value of the benefit of developing the advanced nuclear option is in the order of $3 billion when development takes the form of either an advanced converter or breeder after 2020.*

Among the decisions, there is a slight edge toward developing the breeder in 2030 and no advanced converter.

Second, the economically worst decision among the 25 analyzed here was a program to develop both the advanced converter and the breeder very quickly; any decision to develop both reactor types before 2020 has an estimated net loss of between $3 and $7 billion. A strategy to develop the breeder alone by 2000 also is estimated to have a net loss.

Third, from the estimates made about the value of perfect information, it is probable that the estimates made here would be slightly more favorable to developing advanced nuclear technology if a substantial improvement in knowledge about the uncertain driving variables was attained. In the optimal decision, adding complete certainty about driving variables adds approximately $2 billion to the net benefits. It is somewhat surprising that the value of perfect information is so small. Given this small value, however, *it seems likely that the assessment of the value of advanced nuclear reactors would not be substantially changed by further refinements of the nature of the revelation of the uncertainties.*

Fourth, the standard deviations of the estimates of net benefits tend to cluster around $20 billion. This is an indication of the dispersion of the net benefits over the large number of possible states-of-the-world. Thus, even though the *expected value* of net benefits for the optimal decision approximates $3 billion, the wide variation of net benefits across different realizations produces, on the basis of probabilities used here, a standard deviation of net benefits that is quite high relative to the expected value.

Finally, it should be emphasized that all the estimates in this analysis are made without weighing the environmental costs of advanced nuclear reactor strategies into the calculations.

APPENDIX A: DESCRIPTION OF THE ENERGY MODELS USED BY THE MODELING RESOURCE GROUP (MRG)

Given the broad scope of the CONAES study, the MRG examined and employed only energy models that are capable of projecting demands for and supplies of energy in a substantial number of production and end-use categories through 1990, at a minimum, and if possible through 2010. No model, of course, exactly replicates another in either the breadth of output information it reports or in the way it is structured. Nevertheless, the MRG was able to focus on the following six models which, to varying degrees, satisfied the broad criteria just mentioned (the MRG member and/or associate[s] most closely identified with each model are listed next to its name):

1. Data Resources, Inc. (DRI) Energy Model: Philip Verleger
2. Dynamic Energy Systems Optimization Model (DESOM): William Marcuse (MRG Associate Member)
3. Stanford Research Institute (SRI) Energy Model: Edward Cazalet
4. Federal Energy Administration (FEA) Project Independence Evaluation System (PIES): William Hogan, David Nissen (MRG Associate Member), James Sweeney (MRG Consultant)
5. Energy Technology Assessment (ETA) Model: Alan Manne
6. Nordhaus Simulation Model: William Nordhaus

Each of these models have been designed for a different purpose. Some, such as the DRI and FEA models, were designed principally for short- and medium-term forecasting and policy analysis. The others, to differing degrees, were designed more for longer-range resource and technology assessments. In addition, the DRI and FEA models pay much closer attention to institutional factors, such as the existence of price controls on natural gas and average cost pricing in the electric utility sector, than do the other longer-run models.

The models also differ in the degrees to which data inputs and outputs are aggregated. Some models report greater demand-side detail, both by end-use categories and geographic regions. Others put more emphasis on projecting in detail the use of different energy resources and technologies, nationwide, and separately, by geographic regions.

Finally, the models differ in the degree to which they allow energy demands to be responsive to changes in energy prices. The DRI model, for example, contains a minimum of supply-side information, but incorporates a series of demand elasticities categorized by consuming sector, fuel used, and geographic region. SRI, in its latest version, defines energy demand in terms of nonsubstitutable end-uses, such as space heat and transportation services, which are treated as functions of energy prices. DESOM defines its demands like SRI but does not make them price-sensitive. Interfuel substitution, however, is explicitly modeled through the use of energy process and resource information. Finally, the FEA, ETA, and Nordhaus models, to differing degrees, use both demand and supply-side information to estimate demand and supply responses to variations in energy prices and incomes. A detailed examination of the price and income elasticities of demand incorporated in each of the models appears in Appendix C.

The principal features of each model are described below. This description highlights principally those characteristics pertinent to the uses made of the models in the CONAES study. It is not a critical comparative evaluation. Nor are the order in which the models are discussed and the length of each description indicative of the extent to which the models were used and relied upon in the study. For more in-depth discussion of each of the models, readers are referred to the footnoted citations following the titles of the models.

A.1 THE DRI ENERGY MODEL[1]

This energy model, developed by Philip Verleger, contains approximately 200 econometric equations estimated on annual data for the years 1960 through 1974. The model is designed principally to project energy demands only to the year 1990.

As Figure A.1[2] indicates, the DRI model projects final demand for and consumption of energy in four principal fuel sectors: coal, electricity, petroleum, and natural gas. A submodel of the utility sector, in turn, allocates the generation of electricity between hydroelectric, nuclear, and fossil-fuel generated electricity, bringing the total fuel demand categories to six.

Within each of these fuel-specific demand sectors, the DRI model forecasts demands in 13 geographic regions in the United States and in each of the following four final consuming sectors:

1. Commercial
2. Industrial
3. Residential
4. Transportation

A major advantage of the model, therefore, is the great detail in which it can project a variety of energy demands.

Figure A.1 also illustrated the major inputs that drive the DRI energy-demand projections. The DRI macroeconomic model is the principal driving force behind the regional model that is used to drive energy demands in the principal sectoral models. Projections of energy prices are taken from the FEA model (described below), from judgments about

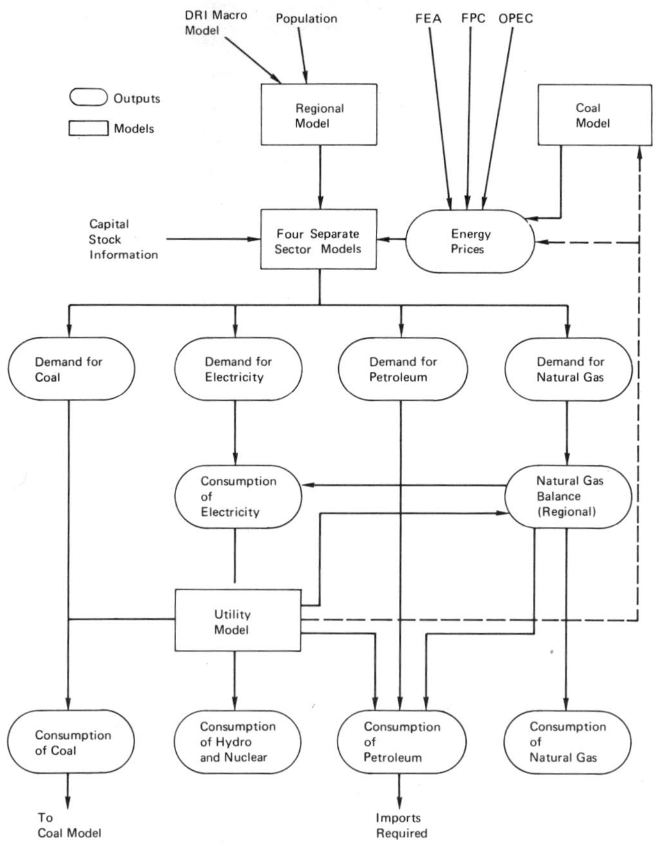

FIGURE A.1 Schematic diagram of DRI energy model.

trends in Federal Power Commission (FPC) regulation of natural gas prices, and from judgmental forecasts of future crude oil prices to be established by the Organization of Petroleum Exporting Countries (OPEC).

Finally, projections of energy-using capital stocks held in each demand sector and region are used as a principal determinant of energy demands. Indeed, a principal feature of the DRI energy model is that energy consumption in some sectors is largely determined by two factors: the kinds of energy-using devices in place and the rates at which such devices are utilized. For example, both the quantity of housing and the desires for particular room temperatures are important determinants of energy demand for space-heating. Analogously, the demand for automobiles of different mileage ratings and the driving intensities of automobile owners will largely determine gasoline consumption. Whereas the stocks of housing and automobiles are then fed into the model exogenously, the intensities of use are based upon behavioral relations estimated from economic data, which indicate how consumers across the United States have responded to income and price movements over the past 15 years. From these relations, the DRI model then projects demand behavior over the 15-year horizon, 1976-1990.

Although the DRI model is designed principally to project only energy demands, short-run constraints on the supplies of certain energy resources are, to a limited extent, explicitly recognized. Thus, regional shortages of natural gas, due to continuation of natural gas price regulation, are forecasted on a judgmental basis, requiring upward adjustments in demands for both petroleum and electricity. Spot excess demands for electricity, however, are left unfilled on the assumption that, in the short run, end-users are unable to substitute another energy source for electricity.

The DRI model has been used for dynamic simulation of energy demands for the period 1965 to 1973. To a large extent the model performed reasonably well, although some weakness appeared in some subsector simulations of commercial, industrial, and electricity demands.[3] In addition, the model has been used to forecast energy demands for 1974, in general, quite successfully.

Up to 1990, then, the DRI energy model (and the FEA model, to be described) possesses the capability of projecting demands for energy in great detail, by fuel type, end-use category, and by geographic region.

A.2 BROOKHAVEN DESOM[4]

In contrast to the DRI energy model, Brookhaven DESOM assumes demands for energy services to be given and then determines the combination of energy resources and technologies that can be used through time to meet those demands at the least cost. It is capable, through dynamic mathematical programming techniques, of making such calculations through the time horizon of the study, 2010 and beyond if desired. It therefore stands as one of the models most useful for the longer-range scenario projections and technology assessments that are the principal tools and objectives of the CONAES study.

DESOM represents a time-phased extension of Brookhaven BESOM, developed by Kenneth Hoffman, to minimize the cost of meeting a set of energy demands in a single period.[5] Both DESOM and BESOM reflect the level of technical detail incorporated in the Reference Energy System (RES) also developed at Brookhaven.[6] BESOM and the RES have been employed in connection with other economic models, such as the University of Illinois input/output model and the Hudson-Jorgenson model. Only DESOM, however, was used in the CONAES study.

The RES is essentially a flow chart that illustrates all of the possible paths through which energy can travel with some losses along the way, from its initial state as primary fuel to its final state as a component of energy demand. It may be employed to display the output of any energy model or projection given sufficient technical detail in the input information. Figure A.2 represents a typical Preference Energy System in which several possible energy paths or "trajectories" are illustrated.[7] One trajectory, for example, is coal extracted from underground, transported to a coal-fired electricity-generating plant, and then used to power an array of electric end-uses.

Many more trajectories outlined in greater regional detail exist, of course, and each consists of a segment of an intermediate energy flow. Associated with each of these flows is an "energy efficiency," which represents the ratio of the energy transmitted to further use or conversion at the end of the flow to that received at the beginning.

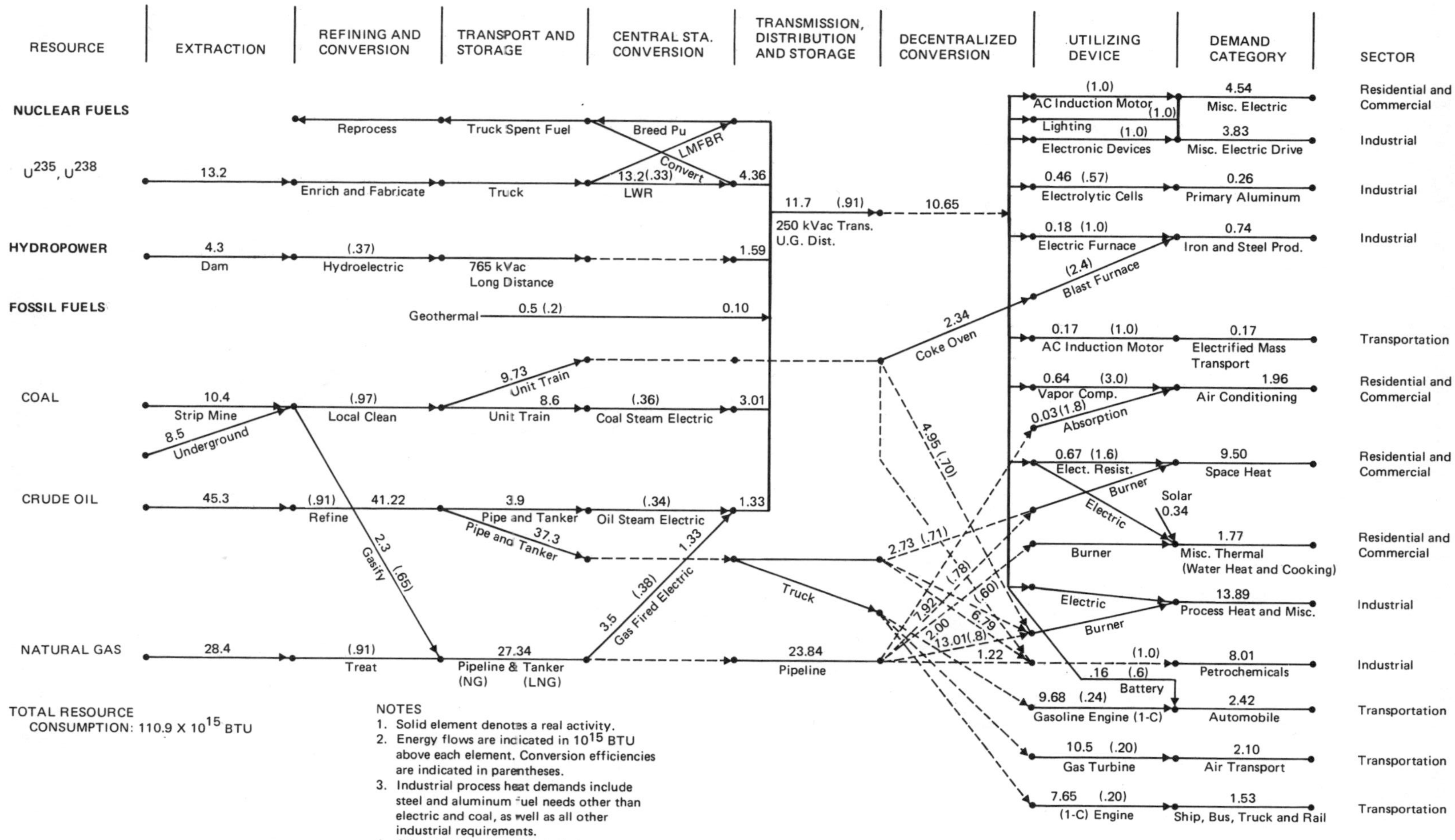

TOTAL RESOURCE CONSUMPTION: 110.9 X 10^{15} BTU

NOTES

1. Solid element denotes a real activity.
2. Energy flows are indicated in 10^{15} BTU above each element. Conversion efficiencies are indicated in parentheses.
3. Industrial process heat demands include steel and aluminum fuel needs other than electric and coal, as well as all other industrial requirements.
4. Oil and gas fired electric includes steam and gas turbine plants.

FIGURE A.2 Reference energy system, year 1985.

Altogether, the Reference Energy System incorporates all of the major extraction, conversion, transmission, and end-use efficiencies associated with each possible trajectory.

DESOM uses the Reference Energy System as its underlying structure, and, given resource supplies at specific extraction costs, capital availability, and/or interperiod limits on capacity expansion, and energy demands in 22 functional end-use categories, DESOM solves for the cost-minimizing fuel and energy technology mix over time. End-use demands over time are generated by applying an income elasticity of 0.75 to the GNP growth rates specified by the MRG (low, base, or high). Substitution between fuels and/or technologies occurs in DESOM automatically as certain fuels and/or technologies become cheaper or more profitable than those they replace. The optimization technique ignores the reduction in total energy demand resulting from higher fuel prices, however, a feature that the authors of DESOM intend to incorporate in future versions of the model.[8]

An approximation to such a procedure can be made with the current version of the model by coupling it directly with the other economic models, such as the Hudson-Jorgenson model, or, with some ingenuity, the end-use energy demands fed into such models can be made consistent with prices through successive iterations.

The DESOM and SRI models also contain explicit representations of load duration curves for electricity generation. By including both seasonal and daily peak-load information, these models are useful for assessing the extent to which base-load generation may be substituted for peak-load generation in the presence of various conversion technologies.

DESOM and SRI can therefore both serve as important tools for policy analysts interested in production and consumption streams of fuel resources over time, selection of alternative research and development activities, or combinations thereof, assessment of policies designed to increase the availability of certain resources, and to deal with the environmental costs and hazards associated with alternative energy policies.

A.3 THE SRI ENERGY MODEL[9]

The SRI energy model, developed under the direction of Edward Cazalet and initially for the use of Gulf Oil Corporation, is similar to Brookhaven DESOM in that it is structured on an energy system or "network" that contains various energy efficiencies, resource availabilities at particular prices, and capacity constraints associated with certain energy technologies. Earlier versions of the SRI model, like DESOM, also treated end-use energy demands as given through time.

In the version of the SRI model used for the final computations in the study, however, the end-use demands were made functions of energy prices. This was primarily accomplished through judgmental rather than econometric techniques, resulting in elasticities for primary energy that are approximately -0.5. (For a discussion of the built-in elasticities in the SRI model, see Appendix C.) The price-sensitive nature of the end-use demands in the SRI model, therefore, is an important difference between the SRI and DESOM models.

Another important difference between the two models is in the computational technique used and resource supply models and certain behavioral relations incorporated in the SRI model which are not features of Brookhaven DESOM. In addition, the SRI energy model contains regional detail on both the demand and supply sides, reporting end-use data for the 9 census regions and output data for 20 supply regions in the United States.

The SRI model generates its projections by simultaneously balancing demands and supplies for energy at each node and at each point in time over its 50-year horizon. Computationally, however, the SRI model does not employ mathematical optimization techniques; the model, therefore, has no "objective function" that is maximized or minimized subject to a set of constraints. Instead, the model assumes that the decentralized actors in the energy system, namely, refineries, power systems, and households, maximize individual objective functions subject to differences in taxes, discount rates, and price differentials. The model then solves for the market equilibrium reached by the entire system at each point in time.

Specifically, the model contains a network of supply and demand functions that reflect how energy supplies and demands at each point in the energy network respond to price changes. Figure A.3 represents a part of the entire energy network upon which the SRI energy model is built. In structure, the SRI energy network parallels the reference energy system developed at Brookhaven, except that, at each "node" in the SRI energy network, the model contains functions that represent market shares captured. The nodes are indicated in the figure by circles for "materials," squares for "processes," and triangles for transportation links. Given the high degree of regional and process detail, there are thousands of such nodes in the SRI energy network, requiring the solution of approximately 100,000 simultaneous nonlinear equations for each simulation over the 50-year time horizon of the model.

The supply curves are generated from the "bottom up," meaning that once supply functions are developed for primary resources, the supply functions at nodes further along the energy trajectory can be developed simply by adding cost information to the supply curve generated at the immediately preceding node.

For example, *cumulative* high-sulfur coal mined in a particular supply region is graphed, on the basis of detailed cost information, as a function of price. When the coal is then shipped via train, the supply curve at the "unit train" node then simply represents the sum of the primary high-sulfur coal-supply function and the supply curve of train transportation services. Supply functions throughout the rest of the energy network are developed in the same fashion.

Demand in the latest version of the SRI is treated very much like supply. At the "top" of the network, namely at the point of final functional end-use, demands are sensitive to energy prices. Demands further down the network, such as space-heating demands for resistance heating or distillate furnace heating, are sensitive, in the model, to the prices of *specific alternative fuels* for the same end-use. This is accomplished

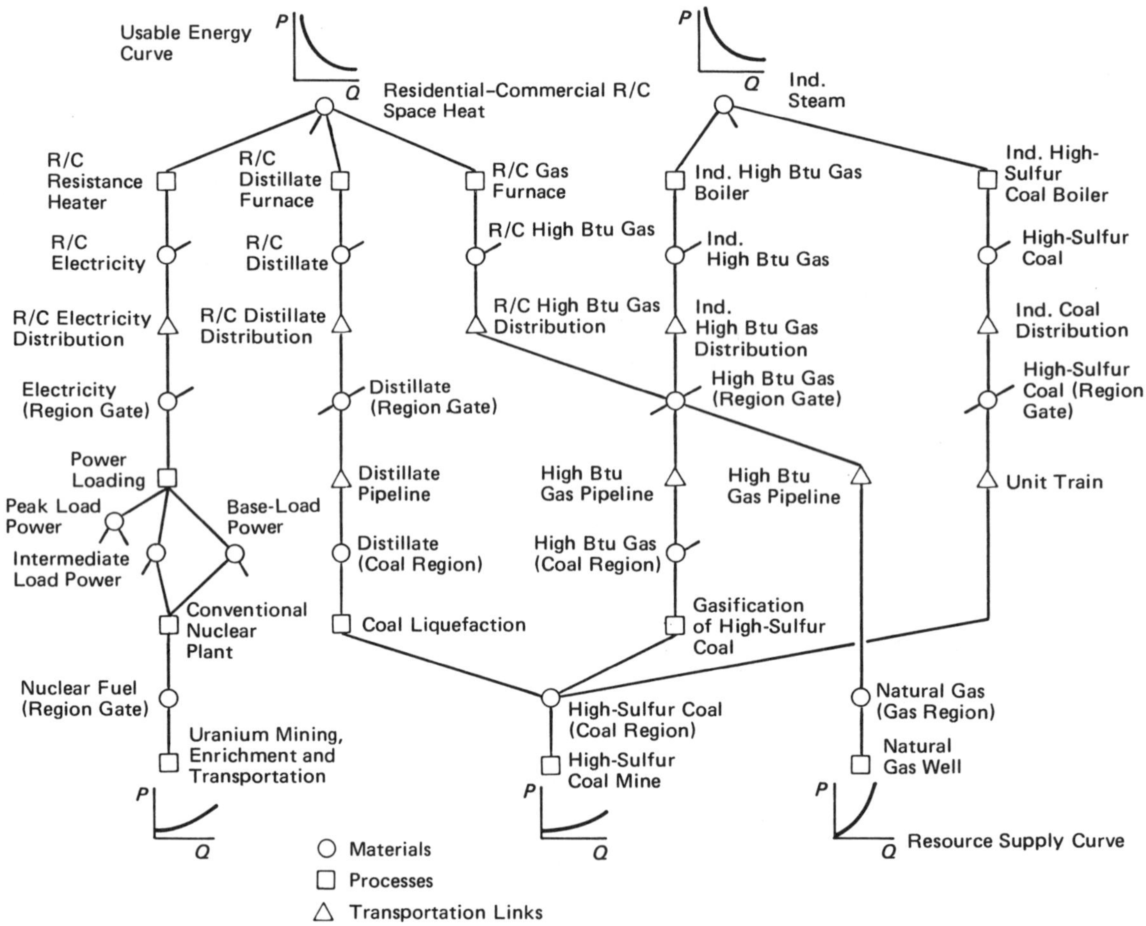

FIGURE A.3 Elements of the energy network.

by specifying the market share of any given fuel as a function of price. Thus, a cheaper fuel will not immediately "drive out" other fuels from the market, as in an optimization model, but rather will capture an increasing share of the market (for a given end-use demand) through time.

The use of such "market share" factors means that the way in which the introduction of new technologies is treated in the SRI model differs from the treatment of the same issue by the mathematical optimization models. In most implementations of mathematical programming models that search for optimal solutions to particular problems, when one technology becomes cheaper than another at some point in time, the optimization procedure will often completely replace the older technology with its new, cheaper alternative. Thus, if at some future time central-station solar power becomes a cheaper way of producing electricity than existing methods, an optimization model might report that, at that time, electric power generated by older methods will fall nearly to zero, while use of solar power will skyrocket.

This "flip-flop" nature of optimization models, described in the foregoing section, can be partially remedied by incorporating constraints on the rate of introduction of capacity for new technologies. Indeed, the MRG has agreed on a set of constraints on the rate of introduction of new technologies that is to apply in all scenario projections. Nevertheless, the "flip-flop" feature of optimization models may, at times, cause the models to give inaccurate estimates of the rates of introduction of new energy alternatives.

A.4 FEA PIES[10]

The Project Independence Evaluation System (PIES) developed under the direction of William Hogan at FEA, with assistance of David Nissen, James Sweeney, and other associates, is one of the most completely integrated demand/supply models of those considered and used by the MRG. Like Brookhaven DESOM, the PIES model is structured upon a Reference Energy System and solves for a minimum cost mix of fuels and energy technologies for sets of final demands. Like the SRI model, the PIES model contains extensive regional detail in its demand and supply sectors. And finally, like the DRI model, PIES incorporates econometrically determined demand equations that project final demands in nine regions for four fuel categories: refined petroleum products, natural gas, electricity, and coal.

The PIES model integrates separate dynamic demand and supply models to describe market equilibria in the energy market in each of 3 years, 1980, 1985, and 1990. Although the equations in the PIES model are dynamic, capital stock formation in the latter years of the simulation period, 1985 and 1990, is not influenced by capital stock formation predicted in the earlier part of the simulation period.

Like the DRI model, the PIES model provides detailed projections of regional and fuel-specific production and consumption through 1990. It also incorporates institutional information not included in the longer-run models. In particular, electric utilities are assumed to price at average (and not marginal) cost in both the PIES and DRI models, and natural gas prices reflect the pattern of price controls prevailing prior to the recent FPC order of September 1976 (whose validity is now being challenged in the courts) and the changing mix of new and old gas contracts. Over the short to intermediate run, these institutional factors can play an important role in any energy simulation; over the long run, however, they should assume less importance, and it is for that reason that the longer-run models used in this study do not incorporate a high degree of institutional detail.

Figure A.4 illustrates the way in which the PIES model integrates its supply and demand features. A macroeconomic forecast from the DRI macroeconomic model is used as the principal economic driving force. The demand model uses the macroeconomic assumptions and conservation options fed into it to generate demand curves. Supply models based on extraction, transformation, and transportation technology and cost information and demand and supply conversion efficiencies generate supply curves. PIES then integrates the supply and demand models through an iterative linear programming technique, which solves for market clearing prices and quantities. The outputs so derived are "optimal" in the

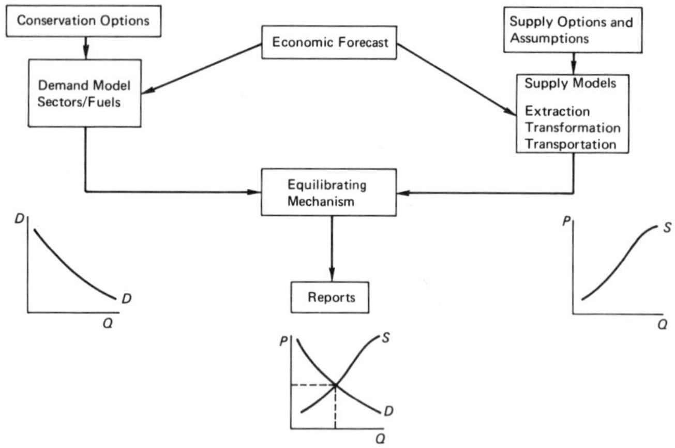

FIGURE A.4 PIES--integrating model framework.

sense that the fuels and technologies used to meet the final demands do so in a way that minimizes the total economic cost of meeting those demands.

A.5 ETA[11]

The ETA models and the Nordhaus model (described below) are independently constructed university-based energy models. They can be characterized as more aggregated equivalents of the PIES model, which extend the time horizon of the latter and for that reason include a wider array of advanced technologies--solar and nuclear.

An important difference between these two models and the PIES effort, however, is that PIES does not contain an explicit objective function that allows policymakers to estimate the gross economic benefits to be gained by introducing new technologies. In addition, both the Manne and Nordhaus models incorporate specific intertemporal constraints; PIES does not.

Specifically, the ETA model approximates market equilibrium in the energy sector through the use of nonlinear optimization techniques. In design, it is identical to Brookhaven DESOM, except that in ETA, energy demands in two broad categories, "electric" and "nonelectric" demands, are sensitive to the prices of electricity and other fuels (including petroleum products, natural gas, and synthetic fuels, but excluding coal). ETA then solves for the fuel and technology mix that minimizes the sum of the costs of conservation, interfuel substitution, and total energy supply.* This model has been used to simulate energy production and use patterns over a 75-year time horizon, well within the 45-year time horizon of the CONAES study.

*An alternative way of viewing the ETA solution is that it maximizes the sum of consumer's plus producer's surplus.

A.6 THE NORDHAUS SIMULATION MODEL[13]

Like ETA, the Nordhaus model is capable of projecting energy production and use patterns over long time periods--90 years for the MRG work. In addition, as in ETA, the model incorporates price-sensitive energy demands, which are met by the mix of fuels and technologies that minimizes total energy costs.

The demand information in the Nordhaus model is more disaggregated than in ETA, however, in that four demand sectors are identified:

1. Residential and commercial
2. Industrial
3. Transportation
4. "Specific" electric

The demand functions used in the Nordhaus model are derived from an extensive econometric study covering seven OECD countries over the period 1955-1972.[14]

An important feature of all of the optimization models, of which ETA and Nordhaus are two examples, is the incorporation of intertemporal constraints on depletable resources. Such constraints apply to the rates of growth of resource production and energy technologies.* The "shadow prices" on these constraints then represent the "economic rents" on these resources.† Economic rent arises because intertemporal profit-maximizing resource owners will not sell their depletable resources at marginal extraction cost; rather, they must earn a return above marginal cost if they are to be induced to sell their resources. Economic rent becomes a larger fraction of the market price at which an alternative fuel becomes cheaper for consumers than the fuel held by the resource owner.

As in all of the energy models discussed thus far, the Nordhaus model solves for market-clearing prices and quantities in only the energy sector, and therefore can be described as a dynamic partial equilibrium model. A strong feature of the model is its detailed treatment of alternative nuclear fuel cycles and capital costs. The model accounts for

*Although neither ETA nor Nordhaus contains an explicit constraint on the introduction of energy-efficiency technologies, the MRG non-price-induced conservation assumption acts in a similar way by gradually phasing in a reduction in demand independent of changes in prices. The non-price-induced conservation assumption is not precisely analogous to the supply side constraints, however, because it includes behavioral and legislative changes as well as improvements in energy-efficiency technologies.

†The SRI model, too, accounts for economic rent, but does so without the use of intertemporal constraints on energy resources. Instead, resource owners are assumed to have perfect foresight and therefore to know the future prices of the resources (the same assumption is made in ETA and Nordhaus, too). They compute economic rents from such knowledge, and are assumed to build such rents into their supply curves.

the physical flows of nuclear materials, and assigns them shadow prices in all scenarios. In addition, the model is capable of choosing a least-cost nuclear reactor configuration (to the extent nuclear reactors are cheaper than alternative means of generating electricity) from among eight reactor types. In this respect, the Nordhaus model, like DESOM, SRI, and ETA, is particularly well-suited to provide insight into some of the principal issues addressed in the CONAES study.

A.7 SUMMARY

Perhaps the best way to visualize the important similarities in and differences between the models, is to examine Table A.1. For each model, the chart indicates whether and, if relevant, in what way, the model features the characteristic itemized in the horizontal rows.

TABLE A.1 A Guide to the Energy Models

	DRI	DESOM	SRI	PIES	ETA	Nordhaus
1. Time horizon date	1990	2025 and beyond	2025	1990	2050	2070 and beyond
2. Final demand definitions, in terms of:						
Energy services	No	Yes	Yes	No	No	No
Fuels	Yes	No	No	Yes	Yes	No
Sectors (residential, industrial, etc.)	Yes	No	Yes	Yes	No	Yes
3. Allowance of own price sensitivity of final energy demands[a]	Yes	No	Yes	Yes	Yes	Yes
4. Incorporation of extensive regional detail	Yes	No	Yes	Yes	No	No
5. Nuclear sector detail	No	Yes	Yes	No	Yes	Yes
6. Incorporation of extensive supply sector technological detail	No	Yes	Yes	Yes	Yes	Yes
7. Incorporation of explicit primary resource supply functions	No	No	Yes	Yes	Yes	Yes
8. Incorporation of load duration curves	No	Yes	Yes	Yes	No	No
9. Other energy process detail	No	Yes	Yes	Yes	Yes	Yes
10. Optimizes over the value of an objective function	No	Yes	No	No	Yes	Yes

[a] "Own" price sensitivity refers to the sensitivity of a particular energy demand to changes in the price of that energy form, e.g., gasoline demand responsiveness to gasoline price changes. "Cross" price sensitivity refers to the sensitivity of a particular energy demand to a change in the price of an alternative energy fuel, e.g., gasoline demand responsiveness to changes in the price of diesel fuel.

NOTES

1. *DRI Energy Outlook: Analyses and Forecasts*, vol. III, no. 2, Data Resources, Inc., February 1976.
2. *Ibid.*, p. 60.
3. *Ibid.*, pp. 75-92.
4. William Marcuse, Lawrence Bodin, Ellen Cherniavsky, and Yasuko Sanborn, "A Dynamic Time Dependent Model for the Analysis of Alternative Energy Policies," *in* K. B. Haley, ed., *Operational Research '75*, Proceedings of the Seventh IFORS International Conference on Operation Research.
5. Kenneth C. Hoffman, "A Uniform Framework for Energy Systems Planning," *in* Milton Searl, ed., *Energy Modelling*, Resources for the Future, March 1973, pp. 108-43.
6. Associated Universities, Inc., *Reference Energy Systems and Resource Data for Use in the Assessment of Energy Technologies*, AET-8, April 1972, revised 1974.
7. See M. Beller, E. A. Cherniavsky, K. C. Hoffman, and R. H. Williamson, "Interfuel Substitution Study--The Role of Electrification," Brookhaven National Laboratory, November 1974.
8. Marcuse, *et al.*, p. 6.
9. *A Comparative Analysis of Methodologies Used in Large Scale Energy Modeling*, Report prepared for the National Science Foundation by the Stanford Research Institute, 1976, Section III.
10. *National Energy Outlook 1976*, FEA, Appendixes A, B, and C.
11. See Alan S. Manne, "ETA: A Model for Energy Technology Assessment," *Bell Journal of Economics*, Autumn 1976.
12. David J. Behling, Jr., Robert Dullien, and Edward Hudson, *The Relationship of Energy Growth to Economic Growth under Alternative Energy Policies*, Brookhaven National Laboratory, March 1976.
13. William D. Nordhaus, "A Simulation Model of the Allocation of Energy Resources and Technology," April 1976 (mimeo), and *The Efficient Use of Energy Resources* (in process). For a description of an earlier version of this model, see William D. Nordhaus, "The Allocation of Energy Resources," *Brookings Papers on Economic Activity*, 1973:3.
14. See William D. Nordhaus, "The Demand for Energy: An International Perspective," in William D. Nordhaus, ed., *Proceedings ot a Workshop on Energy Demand*, 1976. Specifically, the Nordhaus model incorporates the results of econometrically estimated demand functions for energy based on cross-sectional data for energy demands in seven OECD countries over the time period 1955-1972.

APPENDIX B: DEFINITIONS AND EXPLANATIONS OF ASSUMPTIONS UNDERLYING THE VALUES OF CERTAIN DRIVING VARIABLES

This appendix contains detailed definitions and explanations of the assumptions underlying the values of those driving variables that were not fully described in the text. Consequently, several variables that were fully described and explained in the text will not be treated further below.

a. Policy Variables

1. Nuclear Moratoria

The MRG considered a range of possible moratoria on nuclear technologies, including only the Fast Breeder Reactor (FBR), or also all Advanced Converter Reactors (ACR's), or also all other nuclear technologies including new construction starts on the present Light Water Reactor (LWR). The MRG defined in its Base Case a "representative" ACR to be one with a conversion ratio of 0.82 and a capital cost 10 percent greater than that of the LWR.* These specifications might be met either by a high-temperature Gas-Cooled Reactor, Modified Light Water Reactor, or by a Heavy Water Reactor. Rather than "name a winner" at this date, however, the MRG defined this "representative" ACR in terms of objective performance specifications (capital and operating costs and the general description of the fuel cycle) rather than by reactor design.

Of the various types of moratoria that may be imagined, the MRG defined the policies of nuclear moratoria considered in the study to be such as to permit the operation of current nuclear facilities and the completion of those currently under construction. It was estimated by William Hogan of the MRG and Gene Clark of FEA that such policies should result in no new capacity being added after 1983, but allow approximately 70,000 MW^e of installed nuclear capacity in 1983, which would be retired over time (30 years) for normal economic reasons.

*The conversion ratio is defined as ratio of the amount of fertile material converted to fissile material to the amount of fissile material consumed or fissioned in a nuclear reactor.

5. Relaxation of Environmental Standards Applicable to Coal and Nuclear Technologies

This policy is defined by the following set of subpolicies:

1. Removal of all requirements for sulfur removal equipment ("scrubbers") at coal-fired electricity-generating plants.
2. No requirements for sulfur removal from coal.
3. No requirements for land reclamation when strip-mining coal.
4. Repeal of the relevant portions of the Occupational Safety and Health Act (OSHA).
5. Elimination of current automobile emission standards.
6. Reduction of nuclear safety regulations.

It must be emphasized that the MRG expresses no opinions or value judgments on the merits of the above set of policies, and, in their roles as private citizens, MRG members are likely to oppose most or all of them. The MRG specified the policies only in order to determine the economic impact of implementing them. Little numerical description of this simulation is given in the report, however, because it was based on a set of input assumptions that differed from the set of assumptions that was finally used.

William Hogan of the MRG estimated the extent of the reduction in market costs that would be achieved by implementing each of the above subpolicies. These estimates, which in some cases are only "best guesses," are indicated below:

	Cost Reduction (Mills/kWh)	
	Coal-Fired Generation	Nuclear Generation
No scrubbers (No. 1)	-4.5	
No coal preparation (No. 2)	-1.0	
No land reclamation (No. 3)	-0.1	
Repeal OSHA: (No. 4)		
in deep mines	-1.1	
in strip mines	-0.1	
Reduced nuclear safety (No. 6)		-2.0

Hogan also estimates that elimination of automobile emission standards (No. 5) would reduce national gasoline consumption by 140,000 barrels per day (approximately 1 percent of current oil consumption).

Although no careful analysis of the economic effects of these policies was made, it is clear that each of these policies, especially No. 1 and No. 6, have considerable significance. For example, coal emissions abatement techniques add approximately 20 percent to the cost of coal-generated electricity; nuclear safety equipment adds approximately 10 percent to the cost of nuclear-generated electricity. Note that the MRG Base Case assumes that neither No. 1 nor No. 6 becomes policy.

b. Realization Variables

1. Growth Rate of Real GNP

To project GNP growth through 2010, two members of the MRG, William Nordhaus and Philip Verleger, constructed separate projections of the three most important determinants of *potential* GNP growth: (1) population growth, (2) potential labor-force participation rates, and (3) growth in potential output per capita.[3] The projection of population growth was taken from the Census II projections made by the U.S. Bureau of the Census.[4] Projections of the potential labor-force participation rate (defined to be the ratio of the actual labor force to the potential labor force, where the latter is defined as the male and female population between the ages of 18 and 64) assumed an increase from 0.73, its average over the 1970-1975 period, to 0.83 by 2010. Taken together, these assumptions imply a potential labor-force growth rate over the period of 0.94 percent per annum.

The third, and most important component of GNP growth, the growth rate of potential output per worker, can only be projected with a high degree of uncertainty. Projections of this variable require one to confront the question: To what degree can we assume that the rate of technological progress achieved in the past will continue in the future? Economists generally assume that technological progress will continue, although estimates of the future rate of such progress vary widely. The consensus "central value" estimate of Nordhaus and Verleger, agreed upon by the rest of the MRG, was that output per capita will continue to grow at an average of 1.8 percent from 1975 to 2000, and at a rate of 1.5 percent thereafter.

Finally, the Nordhaus and Verleger central value projection of GNP growth envisions that *actual* GNP will complete its recovery to potential GNP by the mid-1980's. Thereafter, actual and potential GNP are assumed to be identical, at levels that would ensure an average unemployment rate of 4.8 percent from 1985 through 2010.

To generate "alternative" values of GNP growth, Nordhaus and Verleger made alternative projections of each of the aforementioned components of potential GNP and of the recovery path of actual GNP toward potential GNP. Of special interest are the assumptions underlying the specification of the low GNP growth path:

1. The potential labor force participation rate is assumed to decline from its average level of 0.73 in the 1950-1975 period to 0.70 in 2010.
2. The average annual unemployment rate through the 1975-2010 period is assumed to be 6 percent.
3. The growth rate of productivity is assumed to decline from 1.57 per annum between 1975 and 1980 to 0 percent annually by 2010.
4. The growth rate of the potential labor force and immigration is assumed to slow to 0.2 percent per annum after 2010.

A high GNP growth path was specified by projecting higher values for each of the above components of GNP. The principal factor that could give rise to growth rates of this magnitude would be continued growth in output per worker at rates slightly in excess of those experienced since World War II.

2. Capital Costs of Electricity-Generating Technologies

All capital cost estimates were approved by a representative of the Supply Panel of the CONAES study. Of special interest are the capital cost figures for solar-produced electricity, which were supplied by the Solar Resource Group of the Supply Panel.

The central value estimate was based upon the projection that a solar thermal technology could produce electricity at a capital cost of $1,200/kW^e at a capacity factor of 0.45. As a 10th percentile (or the "low" estimate), the Solar Resource Group envisioned several technologies as possible candidates (including solar thermal, photovoltaic, and ocean thermal). Without speculating which technology would actually be developed with the most success, Mel Simmons, the chairman of the group, suggested a capital cost estimate of $1,100/kW^e at a capacity factor of 0.90. This approximates the cost figures used by the Solar Resource Group of the Supply Panel for the capital costs of advanced ocean thermal energy conversion systems.

4. U.S. Oil and Gas Resources

As a device for attaining uniformity across all models, the MRG specified a fixed amount of oil and gas resources available at an extraction cost of up to $2/10^6 BTU. No additional resources were assumed to be available at higher extraction costs.

These assumptions are admittedly at variance with the MRG specification of the supply curve for uranium resources, but can nevertheless be justified on other grounds. In particular, the availability of oil and gas resources at extraction costs above $2/10^6 BTU does not appear to have a significant effect on the economic decision of investing in advanced nuclear reactors. This is clearly not the case with uranium, however, where the shape of the supply curve can be critical to the resolution of that issue. Indeed, it was very much for this reason that the MRG spent a considerable amount of time and devoted a great deal of care to the specification of the uranium supply curve (see discussion below). Had the MRG greater time, it would also have been able to specify the oil and gas supply curve in a fashion more consistent with the supply curve specification of uranium resources.

5. U.S. Uranium Resources

The central value estimates were made in two stages. First the group adopted the finding made by the National Uranium Resource Evaluation (NURE) Preliminary Report that 3.7 million tons of U_3O_8 can be expected to be found at a forward cost of less than or equal to $30/lb.[5] To translate this into a price figure, the group assumed that uranium prices were, on the average, marked up 50 percent over forward cost, and hence the 3.7 million tons was assumed by the MRG to be recoverable at a "minimum acceptable price," of less than or equal to $45/lb. For uranium resource estimates below the price of $45, the group also relied on the NURE data.

For uranium resource estimates at prices above $45, the group applied a supply elasticity of 0.76, derived from the supply elasticity implied by the resource estimates in this price range made in an earlier ERDA document, ERDA-1535.[6] While it is true that the absolute quantities of estimates in this price range were lower in ERDA-1535 than those used here, the MRG decided to use the higher NURE figures for resources below $45/lb and then apply the ERDA-1535 supply *elasticity* to estimate resources above $45/lb.

Although lacking an explicit geological foundation, this procedure was adopted in order to incorporate several qualitative considerations that argue for higher *total* resource estimates (i.e., over the entire cost range) than have been made in previous ERDA documents, such as NURE and ERDA-1535. In particular, these considerations were introduced by Dr. Hans Landsberg of Resources for the Future, Inc., in conversation with a few MRG members.* First, he mentioned that resource estimates made in early phases of a mineral industry, such as uranium mining, tend to underestimate the amount of recoverable resources. As examples he mentioned the history of the petroleum and copper industries. Second, he indicated that uranium exploration has thus far been carried out systematically in only one type of geological formation, sandstone, where most low-cost deposits have thus far been found. Any high-cost ores that have been discovered have been limited to by-products of this search for low-cost ores, suggesting that current estimates of high-cost ores may considerably understate available resources in these cost and price ranges. It was partly for this reason that the MRG adopted the supply elasticity extrapolation procedure described above for uranium resource estimates at prices above $45/lb. Finally, the uranium industry has been, until recently, in a long depression, and even now faces considerable uncertainty stemming in large part from the uncertainty surrounding the nuclear industry in general. Such market conditions have significantly retarded exploration.

Consequently, in extrapolating uranium resource estimates to the higher-price ranges in the way that it has, the MRG implicitly has assumed that significant uranium deposits may be found in the future after some market uncertainties have been resolved and the industry has had time to mature.

6. Extraction and Delivery Costs of Coal

The central values are based upon the "1985 regional limitation" scenario simulated by the PIES model,[1] which implies roughly constant coal

*National Academy of Sciences Offices, Washington, D.C., September 29, 1976.

costs through 1990. Such projections, admittedly, rest on the view that future increases in total labor costs will be offset by increases in productivity. Thus, despite short-run problems that may be experienced in the coal extraction industry, it was the considered judgment of the members of the MRG that the high quantity of available low-cost coal resources would allow the extraction and delivery cost to remain constant *in real terms* over the 1975-2010 period.

c. Blend Variables

1. Discount Rates

Conceptually, the MRG worked with the following three discount rates:*

1. The pretax cost of capital in the nonfinancial corporate sector (r_1),
2. the pretax cost of capital in the energy production, conversion, and distribution sector, specifically (r_2), and
3. the posttax discount rate used for private consumption decisions (r_3).

In theory, the MRG required only r_2 and r_3 because the models available to it simulate private decision-making only in the energy sector. In practice, however, since it was quite difficult to obtain reliable estimates of r_2, the MRG assumed the costs of capital in the energy sector specifically and in the nonfinancial corporate sector as a whole to be equal.

Thus, r_1 and r_2 were set at the 1946-1975 average pretax rate of return of 13 percent on all physical capital of nonfinancial corporations, (whether equity- or debt-financed), computed from data in the *Survey of Current Business* by a method used by William Nordhaus.[8] A posttax rate of return to corporate shareholders was then computed by applying the average effective rate of taxation of approximately 45 percent in the nonfinancial corporate sector to r_1. This yielded a net aftertax rate of return to corporate shareholders of approximately 7 percent to which an average personal income tax rate of 15 percent was applied. The result, 6 percent, was then used as r_3, the discount rate for private and social consumption decisions.

Several comments about this procedure are in order. First, because r_3 is computed both in theory and in practice from discount rate data applicable to a broader range of activities than just those in the energy sector, it is independent of the value of a pretax rate applicable specifically to the energy sector (r_2). Therefore, errors in estimating the value of r_2 have no bearing on the rate used for r_3.

Second, because the current corporate tax laws provide an incentive for debt financing by allowing the deductibility of interest costs, the corporate debt/equity ratio will affect the effective rate of taxation.

*This taxonomy was suggested by Martin Feldstein.

The MRG recognized this by using actual historical, rather than statutory, rates of taxation in computing the posttax rate of return.

Finally, the MRG was well aware that the use of the pretax nonfinancial corporate rate of return as the pretax rate of return in the energy sector was clearly a second-best procedure. In particular, because much of the energy sector is regulated in some fashion, there is reason to believe that the rates of return for some segments of the energy sector are lower than those that prevail elsewhere. On the other hand, possibly above-average risks in the past for the unregulated portion of the energy sector and above-average future risks for the sector as a whole argue for a higher rate of return than the nonfinancial corporate rate. Until future research settles this issue, the MRG found no compelling reasons for using a rate in the energy sector, in particular, different from the pretax rate used in the nonfinancial corporate sector generally.

3. Year of Commercial Availability of Various Advanced Electricity-Generating Technologies

For a description of the ACR, see the discussion of "Nuclear Moratoria" in the "Policy Variable" portion of this appendix.

NOTES

1. "Current and Projected Nuclear Capacity," memorandum from William W. Hogan to the Modeling Coordination Group, June 14, 1976.
2. See "Absence of Environmental Restrictions Scenario," memorandum from William W. Hogan to the Modeling Resource Group, July 29, 1976.
3. A detailed breakdown of the potential GNP projection is contained in an internal MRG document, "MRG-9," available upon request from CONAES.
4. *Population Estimates and Projections,* Current Population Reports, Bureau of the Census, ser. P-25, no. 601, October 1975.
5. National Uranium Resource Evaluation (NURE) Preliminary Report (U.S. Energy Research and Development Administration), June 1976, p. 5. This estimate includes the sum of reserves and "Probable," "Possible," and "Speculative" resources. The difference between reserves and the three categories of resources is that reserves are those amounts of ore that, with a high degree of reliability, can be recovered at less than the stated cost; the three categories of resources represent ores which, with declining levels of confidence, can be recovered at less than or equal to the same cost.
6. *Final Environmental Statement, LMFBR,* ERDA-1535, Figure III F-51, vol. 1, Attachment 3.
7. See *National Energy Outlook,* 1976 FEA, Tables IV-X, p. 179. For definitions of the cost estimates of coal provided for electric utilities, see p. F39.
8. See William D. Nordhaus, "The Falling Share of Profits," *Brookings Papers on Economic Activity,* 1974:1, pp. 169-208.

APPENDIX C: INTERMODEL COMPARISON OF BUILT-IN AND IMPLICIT PRICE AND INCOME ELASTICITIES OF DEMAND FOR ENERGY

This appendix documents the elasticities embodied in the models employed in the work of the MRG. In addition an attempt is made to compare the alternative elasticity assumptions of the various models. Because of the complexity of the problem the documentation is preceded by some caveats.

C.1 CAVEATS

Two major problems involved in comparing elasticities between models are their differing points of measurement in the energy system and their differing aggregation of the energy fuels flowing through that system.

Consider first the demand for a single-energy final product. Suppose that in a given model one measures the demand for this product at two different points, point 1 and point 2, respectively, in the sequence of processing, refining, and transportation steps through the energy system. The price per unit of energy content calculated at point 2, say p_2, will exceed the value $p_1 x_1$ of the required number x_1 of energy units at point 1 by the cost, call it c, of supplying one unit of energy to point 2. Then, if one assumes a constant own-price elasticity of demand, η_2, measured at point 2, this will induce a *nonconstant* own-price elasticity of demand, η_1, at point 1. This elasticity of demand is defined as:

$$\eta_1 = \left(\frac{p_2 - c}{p_2} \right) \eta_2. \qquad (1)$$

Thus, when the conversion/transportation cost, c, represents a significant part of the price p_2, the effect of this transformation will be significant. For example, the Nordhaus model's assumed own-price elasticity of -1.28 for energy used in the transportation sector measured at the point of end-use (i.e., axle of the vehicle) seems very much higher than ETA's aggregate own-price elasticity of -0.25 for energy measured at the busbar/refinery gate. When the Nordhaus elasticity is transformed to the refinery gate, however, using the actual prices that are projected by that model for the Base Case in 2010, an own-price elasticity for transportation energy at the refinery gate of only -0.64 is obtained. In the models used by the MRG, the following points of measurement were used: primary energy production, busbar or refinery gate (primary energy minus conversion losses), point of fuel delivery or input to end-use device (busbar energy minus transmission losses), and point of end-use (energy at point of fuel delivery minus end-use conversion losses). Table C.1 indicates the measurement concept used by each model.

The second major problem one encounters in attempting to compare elasticities arises when the energy products they refer to are aggregated differently. For example, suppose one model measured the demand for two energy products individually while another aggregates them. In the disaggregated model the elasticity of aggregate demand with respect to aggregate price (the base-year-quantity-weighted average price, say) depends crucially on the contributions of the increases in the individual energy product prices. If no cross-price elasticities between the two energy products are assumed, the aggregate elasticity could turn out to have any value between the own-price elasticities for the individual fuels, depending on the ratio of the increases in the two product prices *and* on the currently prevailing demand mix. When cross-price elasticities of demand are included, however, not even the negativity of the aggregate elasticity is assured. For example, starting from the Base-Case demand mix and energy prices, predicted by ETA for 2010, the aggregate own-price elasticity would be -0.25 if the prices of electric and nonelectric energy rise proportionally, but -0.35 if the increase in aggregate price is due solely to an increase in the price of nonelectric energy, and +0.09 if it is due solely to an increase in the price of electricity. The reader is urged to keep these caveats in mind both in reading the documentation of the elasticities embodied in the models in the next section, and in assessing the subsequent attempts at elasticity comparisons.

C.2 PRICE AND INCOME ELASTICITIES OF DEMAND FOR ENERGY BUILT INTO THE MODELS USED BY THE MRG

This section describes the definitions of the demand categories and the specification and empirical sources of elasticities for these demand categories (if any) for each of the six models employed by the MRG. Numerical values are presented only for elasticities that are applied under the assumption that they are constant (i.e., invariant to the levels of prices and quantities demanded). Thus, these elasticities are assumed to be independent of the particular scenario or year considered and are neither estimated from simulation runs, nor are they quantity-weighted averages of more disaggregated elasticity data. Reporting such average values in this documentation would not be descriptive of the models, as these values could depend on very different set of assumptions.

After the elasticities for each model are presented, a summary of characteristics of those elasticities is given.

a. Nordhaus Model Elasticities

In the Nordhaus model total energy demands for the United States are projected for each of four consuming sectors. The demand for energy in each sector is measured at the point of end use taking into account end-use conversion losses (and hence is equal to the energy delivered to the sector times the relative efficiency of the given fuel in that sector). Three traditional consuming sectors are represented: residential/commercial, industrial, and transportation. Because many end uses (e.g., air conditioning, electric appliances, and electric motors) can only be satisfied by electricity if one is to avoid very sizable losses in economic efficiency, a fourth demand category termed "specific electric," is also included. A constraint of the programming model then prevents the specific-electric demand of any sector from being satisfied with nonelectric energy. The following table reports the constant long-run elasticities at the point of end-use in each of the four demand sectors specified in the Nordhaus model. These long-run elasticities represent the full extent of the demand response after adjustments have finally been made to any price change. In the Nordhaus model, full adjustment takes place after 20 years.

Sector	Long-Run Per Capita Income Elasticity (At Point of End-Use)	Long-Run Own-Price Elasticity (At Point of End-Use)
Specific-electric	0.92	-0.65
Residential/commercial	1.09	-0.79
	(0.12)*	(0.08)
Industrial	0.76	-0.52
	(0.16)	(0.17)
Transportation	0.81	-1.28
	(0.08)	(0.06)

The elasticities for the residential/commercial, industrial, and transportation sectors were estimated econometrically by Nordhaus (1975, p. 571, Table 23) on the basis of pooled cross-sectional/time-series annual data from seven OECD countries for 1959-1972.† In preparing inputs for the estimation procedure, individual fuel demands in each

*The numbers in parentheses below the elasticity estimates are the estimated standard errors of those estimates.

†The residential/commercial and industrial sector elasticities chosen were estimated with country dummy variables, while the transportation sector elasticities chosen were estimated without country dummy variables, because these were felt to be the most reliable estimates for reasons stated in the reference. The actual number of observations for the individual countries varied from 8 to 14, so that for each regression there were 82 observations, as described on p. 556 of the reference.

sector were multiplied by--and individual fuel prices were divided by--the thermal efficiency of the fuel in that sector to convert them to prices and quantities measured at the point of end-use. These quantities were then summed to get the total sector demand and the corresponding prices were computed at the quantity-weighted averages of the adjusted fuel prices.

The specific-electric elasticities with respect to both income and price were calculated by assuming that the specific-electric portion of each sector's demand (which is by definition independent of the rest of that sector's demand) has an elasticity identical to that for the rest of the demand for that sector. Since the demand categories are themselves assumed to be "independent,"* the specific-electric elasticities can be computed as the quantity-weighted average of the sector elasticities, with the quantities being the amount of specific electricity demanded in the corresponding sector for some base year. Using 1972 as a base year and assuming that in that year in the United States all electric consumption except one-half of the residential/commercial space-heating, water-heating, and cooking electricity consumption in that year was specific-electric consumption, the elasticities for the specific-electric category shown in the table were calculated.

Since the sector demands in the Nordhaus model are assumed to be independent, the cross-elasticities of fuels used in one sector with respect to the prices of fuels used in other sectors are by assumption equal to zero. Interfuel substitution within each sector is, however, modeled endogenously with the fuel (or fuels) that costs the least to use in that sector. The cost of using a particular fuel in a given sector is calculated by adding a fixed charge for distribution and taxes to users in that sector in its production cost and then dividing this total by the relative efficiency of using the fuel in that sector. Consequently, the cross-elasticities between fuels within a particular sector depend on relative prices and are quite discontinuous.

To represent delays in consumer response to the sharp increases in prices that occurred in 1973, it is assumed, in the Nordhaus model, that the elasticities are only -0.05 until 1978 and then rise linearly to their long-run values over the following 20 years. The length of the delay is estimated on the basis of the lifetime of capital goods. Other future price rises are assumed to be perfectly foreseen and, thus, the long-run elasticities prevail without delay after 1998.

b. PIES Model Elasticities

In the PIES model, demands for each fuel within each of the major consuming sectors for each of the nine U.S. census regions are determined. The fuel demand submodels for the residential, commercial, and industrial sectors have a common structure, while the submodels for transportation fuel demands are structured somewhat differently.

*In this paragraph, "independence" is used in the special sense that the indirect utility or demand functions are a function only of own-price and income.

Residential, Commercial, and Industrial Sector Elasticities

Modeling fuel substitution choices in response to price changes is central to the major models in these sectors. The following table shows the fuels whose demands are jointly modeled within each sector.*

Residential Sector	Commercial Sector	Industrial Sector
Electricity	Electricity	Electricity
Natural gas	Natural gas	Natural gas
Distillate oil	Distillate oil	Distillate oil
Kerosene	Residual oil	Residual oil
Liquid gases		Kerosene
		Liquid gases
		Coal (fuel and power)

The demands for these fuels are measured at the point of fuel delivery or of input to the end-use utilizing device (e.g., the shell of the residential dwelling, the gasoline tank, etc.).

They are estimated within each region through a two-step process. First, a total quantity index for the sector and region is estimated, as a function of (1) an aggregate energy price index; (2) the sector's activity level in the region, measured by regional population and personal income (in the residential and commercial sectors) or value added in manufacturing (in the industrial sector); and (3) the lagged value of the total quantity index. The total price and quantity indices in each region are log-linear value weighted averages of the regional prices and quantities, respectively. It is assumed that constant income and price elasticities are appropriate for each of the nine regions. The following elasticities, measured at the point of fuel delivery, were econometrically estimated, using pooled cross-sectional/time-series data for the nine census regions for 1960-1972 (Federal Energy Administration, 1976; Appendix C).

Sector	Long-Run Per Capita Income Elasticity (At Point of Fuel Delivery)	Long-Run Elasticity with Respect to Aggregate Energy Price (At Point of Fuel Delivery)
Residential	1.04	-0.49
Commercial	1.55	-0.30
Industrial	1.00	-0.31

*These fuels represented 90 percent of the demand for fuel in these sectors in 1974. Demands for the other fuels (e.g., demands for metallurgical coal) are assumed to depend on industrial output indices for the industry where the fuel has its primary application.

The second step in the process of estimating regional fuel demands involves determining the ratio of each specific fuel demand to the total quantity index of each region as a function of: (1) the relative price of the fuel in the region (the ratio of the regional fuel price to the regional total energy price index) and (2) the lagged value of the ratio of the specific fuel demand to the total quantity index. The own- and cross-price elasticities for the individual fuels in each sector turn out to be dependent on the value shares of the relevant fuels in that sector.

Transportation Sector Elasticities

Transportation demands in the PIES model are aggregated by region and end-use, but measured, in general, at the point of input to the end-use device. The demands are estimated for the nation as a whole and then disaggregated using regional income and population projections.

Automotive use of gasoline is determined as the ratio of automobile vehicle miles to the average fleet efficiency. The demand for vehicle miles is specified to be a function of per capita income, the cost of vehicle operation (both time and fuel cost), and the unemployment rate. The efficiency of new cars depends upon gasoline price and a measure of average technical efficiency of new cars when standardized for curb weight. As a result, the difference between the price elasticities for "auto vehicle miles" and "auto gasoline demand" (reported in the table directly below) is due to a term that reflects the response of vehicle efficiency to changes in the price of gasoline.

Total fuel demands for trucks and buses in the PIES model are estimated (separately) as functions of vehicle miles and fleet efficiency (without vintaging) and then apportioned between gasoline and diesel fuel on the basis of relative prices, an industrial production index, total vehicle miles, and a measure of the size of the truck stock.

Demand for rail diesel fuel is estimated as a function of rail ton-miles, which in turn is estimated as a function of diesel fuel price and capital expenditures in the railroad industry.

Commercial jet fuel demand is estimated as a function of air passenger miles, adjusted by an airplane load factor to account for excess capacity. Air passenger miles are estimated as a function of operating cost (a part of which is the price of jet fuel) and per capita income. Total jet fuel consumption also includes military naphtha-type jet fuel demands that are specified exogenously.

The constant elasticities of the major fuels in the PIES transportation submodels are as shown in the following table.

Demand	Long-Run Income Elasticity	Long-Run Elasticity with Respect to Specific Fuel Price
Auto vehicle miles	0.976	-0.150
Auto gasoline demand	0.98	-0.720
Airline passenger miles	1.457	-0.245

Truck fuel demand	1.740	-0.545
Bus fuel demand	0.285	-0.475
Rail diesel fuel demand	0.144	-0.368

These elasticities were econometrically estimated using national data for 1960-1972 (FEA, 1976, Appendix C; Sweeney, 1977).

Modeling the Adjustment to Equilibrium

Price trajectories are constructed in PIES by assuming prices rise (or fall) linearly from their 1974 levels to the equilibrium end-point prices from a comparable PIES run in 1980, and then remain constant through 1990. Thus, since the PIES demand model is dynamically specified, the elasticities used in the model for a particular year will, in general, be less than the long-run elasticities shown here, to reflect the lagged responses to these price movements.

c. ETA Model Elasticities

In ETA energy demands are divided into two categories representing total U.S. demands for electric and nonelectric* energy. These energy demands are measured *before* final delivery to the end-user (i.e., at the busbar for electricity demands and at the refinery gate for nonelectric demands). Constant long-run own- and cross-price elasticities and constant long-run income elasticities are specified for these two categories as given in the following table.

Sector	Long-Run Per Capita Income Elasticities	Long-Run Price Elasticities	
		Own	Cross
Electric	1.0	-0.75	0.50
Nonelectric	1.0	-0.50	0.25
Total	1.0	-0.25	--

ETA postulates (Manne, 1976) that market demand curves are the outcome of a process in which the U.S. economy maximizes its welfare through allocating its expenditure budget optimally between energy and other items. The model also postulates that the amount spent on energy is sufficiently small so that it does not affect the marginal utility of consumers' expenditure on items other than energy. With a unit

*Nonelectric energy as defined in ETA includes only gaseous and liquid fuels and, therefore, excludes coal required for direct uses. Demands for coal for direct uses (that is, not used for liquefaction or electricity generation) are specified exogenously in ETA.

elasticity of substitution in consumption* between electric and nonelectric energy, the maximand during any one time period may be expressed as:

Benefits Obtained from Consumption of Electric and Nonelectric Energy (Unit Elasticity of Substitution)		Benefits Obtained from Expenditures on All Other Items--Given a Total Budget of B
$aq_e^{b_e} q_{ne}^{b_{ne}}$	+	$[B - c(q_e, q_{ne})]$

Where q_e and q_{ne} are the quantities of electric and nonelectric energy demanded, $c(q_e, q_{ne})$ is the cost of supplying these quantities, and a, b_e, and b_{ne} are constants.

It will be optimal for energy consumers to purchase quantities q_e and q_{ne} such that the partial derivatives of the objective function are equal to zero. This leads to the derivation of demand functions with constant own- and cross-price elasticities, which are in turn functions of the parameters b_e and b_{ne}, as shown in the following table.

Long-Run Price Elasticities

Sector	Own	Cross
Electric	$\frac{b_{ne} - 1}{1 - b_e - b_{ne}}$	$\frac{-b_{ne}}{1 - b_e - b_{ne}}$
Nonelectric	$\frac{b_e - 1}{1 - b_e - b_{ne}}$	$\frac{-b_e}{1 - b_e - b_{ne}}$
Total	$\frac{1}{b_e + b_{ne} - 1}$	

Since a unit elasticity of substitution in consumption between electric and nonelectric energy is assumed in the model, the ratio of the two parameters (b_e/b_{ne}) must be invariant with respect to variations in the quantities demanded.

A value of one-half for this ratio is inferred from historical data by Schmitt *et al.* (1976). They then show analytically that if (1) the historical behavior of q_e and q_{ne} and their respective prices p_e and p_{ne} can be represented by exponential functions of time, and if (2) in a moving equilibrium the prices of both electric and nonelectric energy were to

*The assumption of a unit elasticity of substitution implies that a 1 percent increase in the ratio of the price of electric energy to the price of nonelectric energy will result in a 1 percent increase in the ratio of the demand for nonelectric energy to that for electric energy.

remain constant, then the growth rates of q_e and q_{ne} would each be equal to the growth rate of GNP. Therefore (q_e/q_{ne}) equals the ratio of total expenditures on electric energy to total expenditures on nonelectric energy, which has historically been roughly one-half. Manne (1976, pp. 394-396) has also inferred this same value from the results of the interfuel substitution study of Hoffman and Cherniavsky (1974). He noted that the electric/nonelectric trade-offs predicted for 2000 are approximately linear when plotted on a log-log graph--and hence consistent with the assumption of unitary elasticity of substitution--with the slope of the trade-off line (b_e/b_{ne}) being approximately one-half.

The absolute level of the parameters b_e and b_{ne} (-1 for b_e and -2 for b_{ne}) were chosen so that the price elasticity of the demand for energy as a whole (varying the prices of both electric and nonelectric energy in the same proportion) is -0.25. This numerical value appears roughly consistent with the elasticities implicit in the BTU tax estimates of Hudson and Jorgenson (1974, pp. 501-509).

The constant a is selected so that aggregate energy demand would grow at the same rate as the GNP if energy prices remain at their low 1970 levels. That is, the long-run per capita income elasticities for each of the two forms of energy are assumed to be 1. This is roughly consistent with the estimated values for this parameter in the literature (e.g., the values assumed in the other models employed in this study: see Table C.3).

To handle the distinction between short- and long-run price elasticities of demand, a "putty-clay" substitution model is employed, with the 1970 initial energy-using capital stock being replaced at the rate of 3 percent per year.

d. DRI Model Elasticities

DRI's econometric demand model is more structured than those employed in the ETA, Nordhaus, or FEA models, in that it emphasizes other determinants of energy demand besides prices and income (e.g., capital stocks, commercial employment, etc.).

In the DRI model, energy demands are generally forecast by fuel for each of four consuming sectors: residential, commercial, industrial, and transportation. Coal and petroleum demands for the residential and commercial sectors, however, are forecast on a combined residential/commercial sector basis. These demands are measured, in general, at the point of input to the end-use device. Most of these sectoral fuel demands are forecast for 13 regions that are defined so as to encompass completely both the census and PAD (Petroleum Administration for Defense) regional definition systems. However, some demands are forecast on a national basis, as indicated in the following table.

Sector	Fuel	Elec-tricity	Natural Gas	Petroleum	Coal
Residential		R	R		
Commercial		R	R		
Residential/commercial				N	N
Industrial		R	R	N	N
Transportation				N[a]	

[a]Except gasoline which is treated on a regional basis.
NOTE: R = regional basis; N = national basis.

The constant long-run own-price elasticities built into the DRI model are given in the following table.

Sector	Fuel	Elec-tricity	Natural Gas	Petroleum	Coal
Residential		-0.33	-0.57	-0.186 (Residential and Commercial)	--
Commercial		-0.58	-0.49		--
Industrial[a]		-0.62	-0.99	-1.17	-0.76
Transportation				-0.16[b]	

[a]The industrial price elasticities shown here are indicative of substitution effects only. The aggregate own-price elasticity for industrial energy is about -0.5.

[b]Elasticity for automobile vehicle miles.

The constant long-run income elasticities embodied in the DRI model are given in the following table.

Sector	Long-Run Per Capita Income Elasticity
Residential	1.0
Commercial	0.8
Industrial	1.0
Transportation	0.7

The regional elasticities were estimated using pooled cross-section/time-series data from the 13 regions and were generally estimated over the time period 1961-1974 (Data Resources, Incorporated, 1976). The national elasticities were estimated using national time-series data over the same interval. The estimation of these elasticities is now discussed by sector.

Residential Sector Demand Estimation

Residential electricity demand is estimated as a function of price, real disposable income per capita, and a variable equal to the housing stock, times the sum of the saturation rates for individual electrical appliances, times the average kWh use for each appliance in 1 year.

Residential natural gas demand is estimated as a function of price, disposable income per capita, and a measure equal to the housing stock, times the sum of the saturation rates for gas appliances, times the average BTU use for each appliance in 1 year.

Commercial Sector Demand Estimation

Commercial electricity demand is estimated as a function of price, disposable income per capita, and commercial employment.

Commercial natural gas demand is estimated as a function of price, commercial sector employment, and lagged demand.

Residential/Commercial Sector Demand Estimation

Residential/commercial coal demands are exogenously specified. Residential/commercial petroleum demand is estimated as a function of price, the housing stock, and heating degree days.

Industrial Sector Demand Estimation

Electricity demand in the industrial sector is estimated as a function of price and a measure called the potential demand for electricity. Potential demand for a fuel is estimated as a function of the projected regional outputs for the 19 Census of Manufacturing two-digit industries, times flexible input-output coefficients for each industry. The input-output coefficients are in turn estimated as functions of the price of the fuel relative to other fuels. Natural gas demand in the industrial sector is estimated as a function of price and the potential demand for natural gas.

Total industrial fuel demand is estimated as a function of total potential demand and a fuel price index. Total potential demand is the sum of (1) the regionally estimated potential electricity and natural gas demands over all regions, and (2) the nationally estimated coal and petroleum demands. The total demand less electricity and natural gas demands is apportioned between coal and petroleum as a function of their relative potential demands and a measure of the impact of environmental constraints on the split.

Transportation Section Demands Estimation

In the DRI model transportation demands are estimated partially by end use and partially by fuel. All transportation demands except automobile gasoline demands were estimated regionally as a function of the ratio of vehicle miles demanded, divided by the weighted miles per gallon characteristic of the stock of automobiles. The demand for vehicle miles is estimated as a function of vehicle registrations per capita, disposable income per capita, and the price of gasoline.

Separate equations were used to estimate military and commercial jet fuel demands. Military jet fuel demand was estimated as a function of federal government expenditures, while commercial jet fuel demand was estimated as a function of airline seat miles, which is assumed equal to revenue passenger miles, divided by the load factor, and time. On-highway use of distillate fuel was estimated as a function of truck registrations. Other distillate fuel, residual fuel, and liquefied gas demands for transportation were all estimated as a function of generated output in the transportation industries.

e. DESOM Elasticities

In DESOM total U.S. energy demands are measured at the point of end-use and aggregated according to the following energy end-use demand categories (Marcuse, *et al.*, 1975, p. 27):

End-Use Demand Category	Sector
Miscellaneous peak-load electricity	Residential/Commercial and Industrial
Miscellaneous intermediate-load electricity	Residential/Commercial and Industrial
Miscellaneous base-load electricity	Residential/Commercial and Industrial
Miscellaneous low-temperature thermal	Industrial
Miscellaneous medium-temperature thermal	Industrial
Miscellaneous high-temperature thermal	Industrial
Iron-ore reduction	Industrial
Petrochemicals	Industrial
Space heat	Residential/Commercial
Air conditioning	Residential/Commercial
Water heat	Residential/Commercial
Air transport	Transportation
Truck, bus	Transportation
Rail	Transportation
Automobile	Transportation

In the DESOM model it is assumed that the own-price elasticity of demands for each end-use category is zero. For example, total vehicle miles driven do not change if gasoline price changes, nor does total space heat demand change when either fuel oil or electricity prices increase. The model treats end-use demands in this fashion because its principal purpose is to solve for the supply mix that meets a given set of end-use demands in a least-cost fashion. Interfuel substitution, on the other hand, occurs automatically in response to price changes as part of the optimization procedure employed in the model. Since the least-cost fuel available is always selected to satisfy a given end-use demand, the cross-elasticities of demand between fuels within a given end-use demand category are price dependent and discontinuous. The end-use demands themselves are assumed to be independent.

Each of the end-use demands is explicitly responsive to changes in aggregate income, however. The income elasticity built into the model for all end-use demands is 0.75, a figure somewhat lower than that used in some of the other models, but supported, in part, by the work of Hudson and Jorgenson (1974).

f. SRI Model Elasticities

In the SRI model regional energy demands are measured at the point of end-use and aggregated according to the following energy end-use demand categories (Synfuels report, 1975):

End-Use Demand Category	Sector
Auto	Transportation
Bus/truck	Transportation
Air	Transportation
Marine	Transportation
Space heat	Residential/Commercial
Miscellaneous heat	Residential/Commercial
Electromechanical	Residential/Commercial
Process steam	Industrial
Direct heat	Industrial
Electromechanical	Industrial
Naphtha feedstock	Industrial
Gas feedstock	Industrial
Coal feedstock	Industrial

In the SRI model it is assumed that the own-price elasticity of demand for each end-use category is -0.7, with the exception of the own-price elasticity of the demand for automobile vehicle miles, which is assumed to be -0.38 to reflect the fixed costs of auto travel included in the model data base.* Cross-elasticities between the fuels used to satisfy a particular end-use demand are represented through the use of econometrically estimated market share curves in the SRI model and are, thus, a function of relative prices. These cross-elasticities are also lagged through the use of judgmentally estimated behavioral response curves.

There is an important difference between the application of the own-price elasticities built into the SRI model and those built into the ETA and Nordhaus models. In these other two models the price elasticity

*As we would predict using equation (1).

applies to the entire variation between the levels of demand that would occur had prices remained at their assumed 1970 equilibrium levels and the calculated equilibrium values. In the SRI model, on the other hand, they apply only to variations in price from the Base-Case equilibrium prices. The end-use demands for the Base Case related to changes in aggregate income and price, but no consistently applied income elasticity has, apparently, been reported for the SRI model.

g. Summary Comparison of Demand Elasticities

Table C.1 is a summary comparison of the characteristics of the demand categories and demand elasticities built into the various models employed by the MRG.

C.3 INTERMODEL ELASTICITY COMPARISONS

This section suggests and attempts to implement a method for comparing the elasticities documented in the previous section. The method, while requiring some output data from the models, is basically analytical in nature.

Analytical Comparisons of Built-in Elasticities

Intermodel elasticity comparisons can be made by transforming the built-in elasticities reported in Section C.2 to a common point of measurement using equation (1) from Section C.1 and then aggregating the component elasticities. Since these aggregate elasticities are computed as quantity-weighted averages, they are estimated on the basis of proportional price increases for the individual fuels.

Perhaps the easiest sets of elasticities to compare are those embodied in the FEA and Nordhaus models, since they are aggregated similarly and measured at nearly the same point in the energy-transformation system. Recall that in the FEA model, own-price elasticities are measured at the point of input to the end-use device, while in the Nordhaus model own-price elasticities are measured in terms of end-use energy (i.e., after end-use conversion losses have been taken into account). However, since end-use conversion costs* are measured relative to those of currently existing end-use devices in the Nordhaus model, and because in the Base Case no new end-use devices (e.g., electric cars) are introduced before 2010, the end-use transformation costs are virtually zero between 1975 and 2010.

Although the FEA model does not define a "specific-electric" demand category, for comparison purposes we will use the Nordhaus procedure for

*These are the costs of purchasing and operating the equipment required to utilize the fuel (e.g., the space-heating equipment).

TABLE C.1 Comparison of the Characteristics of the Demand Categories and Elasticities Embodied in the Models Employed by the CONAES Modeling Resource Group

Model	Spatial Aggregation	Product Aggregation	Point of Measurement[a]	Own-Price Elasticities Considered	Cross-Price Elasticities Considered	Modeling Technique for Cross-Elasticities
DESOM	U.S. total	End-use	Point of end-use	No	Between fuels within end-uses	Math programming
DRI	13 U.S. (census/PAD) regions[b]	Fuel by consuming sector[c]	Input to end-use device[d]	Yes	Between fuels within sectors[e]	Econometrics
ETA	U.S. total	Electric Energy Nonelectric energy[f]	Busbar Refinery gate	Yes	Between electric and nonelectric energy	Econometrics
Nordhaus	U.S. total	Consuming sector totals	Point of end-use	Yes	Between fuels within sectors	Math programming
PIES	9 U.S. (census) regions	Fuel by consuming sector[c]	Input to end-use device	Yes	Between fuels within sectors	Math programming and econometrics
SRI	8 U.S. regions	End-use	Point of end-use	Yes	Between fuels within end-uses	Math programming and econometrics

[a]Refers to the point of measurement of energy demands in physical units (e.g., BTU's), not at the terminus of the energy-supply trajectory cost parameters. For example, although both the DESOM and SRI models measure energy demands at the point of end-use, the SRI model includes the costs of utilizing devices in computing energy system equilibria, while DESOM does not.

[b]Except for coal and petroleum demands, which are estimated on a national basis.

[c]Except for demands within the transportation sector, which are aggregated by end-use.

[d]Except for industrial demands, which are measured at the point of end-use.

[e]Cross-elasticities are included only for industrial fuels.

[f]Excludes coal required for direct burning (i.e., coal demands other than for electricity generation or synthetic fuels). Coal demands for direct use are assumed to grow at 2 percent per annum in ETA.

TABLE C.2 Comparison of the Own-Price Elasticities for Delivered Fuels in the FEA and Nordhaus Models

	Own-Price Elasticity	
Consuming Sector	FEA Model	Nordhaus Model
Transportation	-0.72	-1.27
Residential/commercial	-0.40	-0.79
Industrial	-0.31	-0.52
Specific-electric	-0.45	-0.65
Aggregate	-0.51	-0.78

constructing the elasticity for such a demand category based on the FEA component sector elasticities.* Additionally, it is assumed that the residential/commercial aggregate elasticity for the FEA model is the average of the elasticities actually specified for the two individual sectors. These assumptions result in uniformly greater own-price elasticities for the Nordhaus model than for the FEA model, as shown in Table C.2. In fact, using the 2010 Base-Case demands projected by the Nordhaus model as weights, an aggregate price elasticity is estimated for the Nordhaus model that exceeds that for the FEA model by approximately 50 percent.

The next step is to convert these elasticities to the point at which ETA's demands are measured (the busbar for electric energy, the refinery gate for nonelectric energy). Fortunately, it turns out that for the Nordhaus model in the Base Case by 2010 each consuming sector's demand is met by a single fuel. Thus, using equation (1) and then aggregating transforms the individual sector demands back to the appropriate point. Although the FEA model actually includes (continuous) price elasticities between fuels within each consuming sector, it is assumed here that the elasticities in Table C.2 apply.

Recall now that the elasticity transformation in equation (1) depends on the prevailing prices. An obvious choice of interesting prices to use are the Base-Case projections made by a given model for a particular year. The Nordhaus Base-Case price projections for 2010 (see Section III.7) can serve this purpose. However, these price levels are ones that suppress demands enough to equal the supplies that are forthcoming at those prices (as shown in Figure C.1). The calculation of the degree of reduction of demand in response to price increases, then, starts with a price level, labelled "reference price" in Figure C.1, that approximates those that prevailed in the 1970-1972 period when the energy system was close to long-run equilibrium. (Recall that with a constant demand elasticity the demand curve would appear as a straight line on a double-log scale.) Hence, using this approach, the estimated aggregate

*Recall these consuming sectors are industrial, commercial, residential, and transportation in the FEA model.

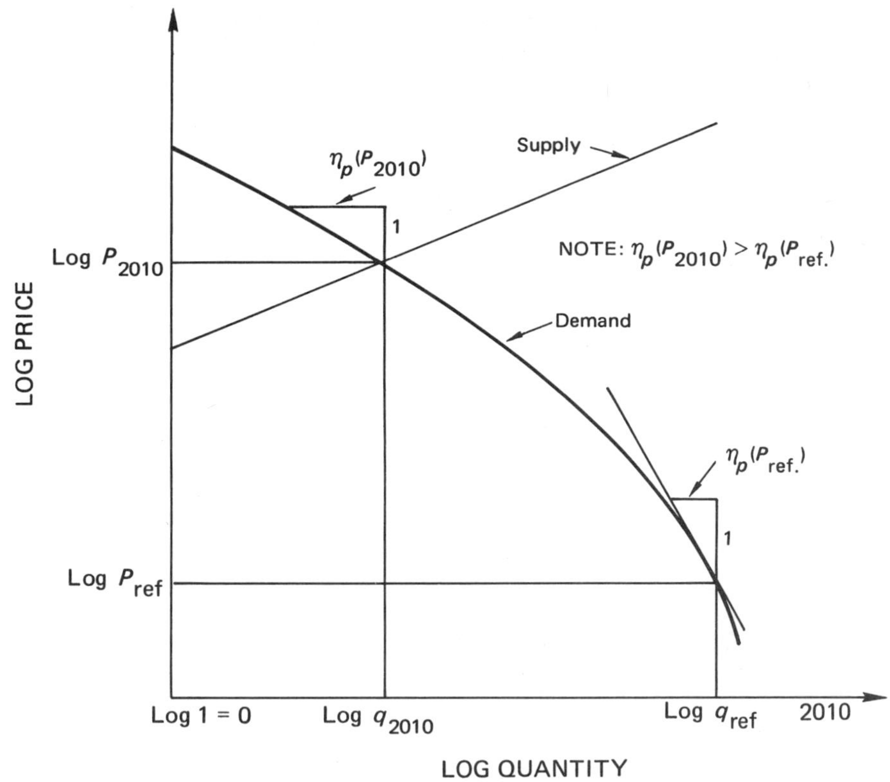

FIGURE C.1 Conceptual illustration of the calculation of equilibrium prices and quantities at the busbar/refinery gate in the FEA and Nordhaus models. NOTE: (1) For ETA this curve would be a straight line; (2) the slopes at the intercept with the ordinate axis would be the built-in own-price elasticities at their actual point of measurement in the FEA and Nordhaus models.

price elasticity for total energy before transmission and distribution increases from -0.31 to -0.49 in the Nordhaus model as prices increase from the reference level to the calculated equilibrium level for 2010.

Table C.2 indicates that the aggregate price elasticity estimated for the Nordhaus model is greater than ETA's theoretical aggregate price elasticity, which appears to be roughly equal to that estimated for the FEA model. As prices rise (proportionally) beyond the projected Base-Case projections for 2010, however, both the Nordhaus and FEA model aggregate price elasticities will exceed that built into the ETA model.

The aggregate income elasticities do not depend on where they are measured in the system, but are dependent on the quantities of the individual energy products demanded as shown in Table C.3. Here the ETA model's built-in per capita aggregate income elasticity of 1 can be compared with the Nordhaus estimate, which is about 10 percent lower, and the FEA estimate, which is about 10 percent higher.

TABLE C.3 Comparison of the Estimated Elasticities for Aggregate Energy at the Busbar/Refinery Gate Evaluated for 2010 (Reference and Base-Case Prices and Quantities Projected by the Nordhaus Model)

	ETA Model	Nordhaus Model		FEA Model	
	For Any Prices and Quantities	For Reference Prices and Resulting 2010 Quantities	For 2010 Base Case Prices and Quantities	For Reference Prices and Resulting 2010 Quantities	For 2010 Base Case Prices and Quantities
Estimated aggregate price elasticity	-0.25	-0.31	-0.49	-0.21	-0.32
Estimated aggregate income elasticity	1.00	0.91	0.89	1.10	1.08

NOTE: "Reference" prices are approximately those that prevailed during the 1970-1972 period.

TABLE C.4 Estimated Price Elasticities of Demand for Electric and Nonelectric Energy at the Busbar/Refinery Gate in the ETA and Nordhaus Models

	With Respect to the Price of:					
	Electric Energy			Nonelectric Energy		
		Nordhaus Model			Nordhaus Model	
Elasticity of the Demand for:	ETA Model	For 2010 Reference Prices and Demands	For 2010 Base Case Prices and Demands	ETA Model	For 2010 Reference Prices and Demands	For 2010 Base Case Prices and Demands
Electric energy	-0.75	-0.32	-0.41	0.50	0	0
Nonelectric energy	0.25	0	0	-0.50	-0.31	-0.51

Finally, with regard to the mix of demand for electric and nonelectric energy in the Nordhaus model, the cross-price elasticity between electric and nonelectric energy is virtually zero for the Base-Case price projections from that model for 2010. To supplement this observation, Table C.4 presents estimates for the own-price elasticities for electric and nonelectric energy. The results show that in Nordhaus, the estimated own-price elasticity for nonelectric energy is slightly in excess of that for electric energy.

REFERENCES

Data Resources, Incorporated, "The Data Resources Energy Model: Model Description," July 1976.

Federal Energy Administration, "National Energy Outlook," U.S. Government Printing Office, 1976.

Hoffman, K. C., and E. A. Cherniavsky, "Interfuel Substitution, the Role of Electrification," Brookhaven National Laboratory, June 1974.

Hudson, E. A., and D. W. Jorgenson, "U.S. Energy Policy and Economic Growth 1975-2000," *Bell Journal of Economics and Management Science*, Autumn 1974.

Manne, A. S., "ETA: A Model for Energy Technology Assessment," *Bell Journal of Economics*, Autumn 1976.

Marcuse, W., L. Bodin, E. Cherniavsky, and Y. Sanborn, "A Dynamic Time Dependent Model for the Analysis of Alternative Energy Policies," Brookhaven National Laboratory, 1975.

Nordhaus, W. D., "The Demand for Energy: An International Perspective," *in* W. D. Nordhaus, ed., *Proceedings of a Workshop on Energy Demand*, May 22-23, 1975, International Institute for Applied Systems Analysis, Report CP-76-1, 1976.

Schmitt, R. W., D. J. BenDaniel, P. J. Stewart, and A. S. Manne, "An Econometric Analysis of Energy Over the Next 75 Years," paper presented at the Joint Power Generation Conference, Buffalo, N.Y., September 19-23, 1976.

Sweeney, J. L., "The Demand for Gasoline: A Vintage Capital Stock Approach," draft report, Department of Engineering--Economic Systems, Stanford University, 1977.

Synfuels Interagency Task Force, "Recommendations for a Synthetic Fuels Commercialization Program," U.S. Government Printing Office, November 1975.

APPENDIX D: AN APPROACH TO THE PROBLEM OF COMPARING INTERMODEL DIFFERENCES

This appendix develops an approach to the problem of explaining differences in the total primary energy consumption projections made by the models. The approach begins by comparing projections made by ETA and Nordhaus using the analytical form of ETA's objective function and assuming flat energy supply curves for electric and nonelectric energy over relevant ranges in each scenario and time period.

At the outset, it is important to distinguish between the concepts of "primary" and "secondary" energy. Primary energy is counted by measuring the energy content (in Quads) of the primary energy resources--oil, gas, and coal. Secondary energy refers to the energy content (in Quads) of energy after conversion losses--in synthetic fuel production, electricity generation, and the like. In the procedure used here, differences in primary energy projections between the models due to differences in income and price elasticities of demand (see Table D.1) are computed after conversion losses are subtracted, that is, at the point where secondary energy is measured. Other sources of the difference between primary energy projections of the two models (see Table D.2) are then treated directly at the point where primary energy is measured.

The differences in model projections of secondary energy consumption in 2010 can be largely explained in terms of: (1) differing income elasticities for aggregate energy demand, and (2) differing price elasticities for aggregate energy demand. Figure D.1 shows a schematic representation of the effects of differing assumptions about the values of these parameters and Table D.1 shows explicitly the implications of these differing assumptions.

Table D.1, for example, indicates that 19.3 Quads of the total 49.1-Quad difference in the projections of primary energy consumption are due solely to the differences in price and income elasticities. The income elasticity difference between the two models (1.0 [ETA] - 0.89 [Nordhaus] = 0.11) accounts for 13 Quads and the difference in price elasticities accounts for the remaining 6.3 Quads. As the table indicates, both figures are computed by excluding the 9 Quads of directly burned coal assumed by ETA for the year 2010, and by using the ratio of primary to secondary energy of 1.3 in the Nordhaus model.

Differences in price and income elasticities account for less than half of the total 49.1-Quad difference to be explained. However, the remaining difference of 29.8 Quads (49.1 - 19.3) can be attributed to:

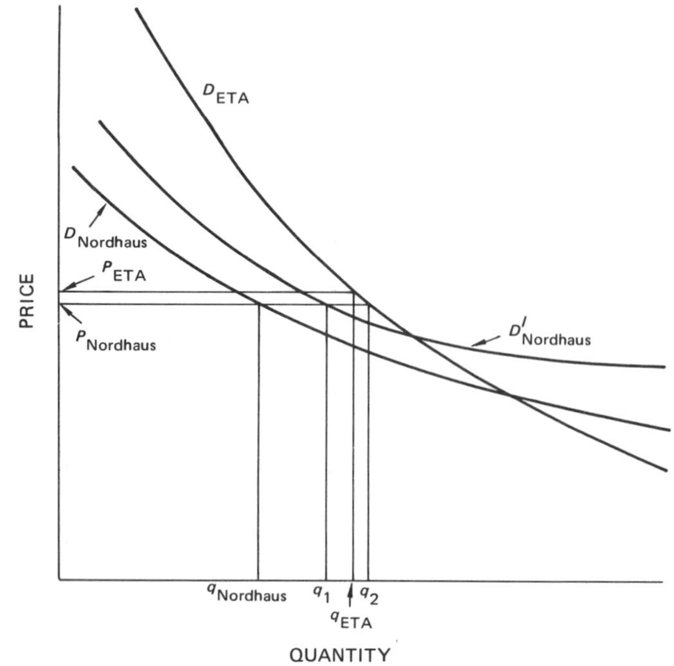

FIGURE D.1 Schematic representation of the aggregate demand specifications of the ETA and Nordhaus models in 2010 for the Base-Case scenario.

1. differences in price projections;
2. differences in synthetic fuel conversion losses;
3. differences in electricity-generation conversion losses attributable to differing cross-elasticities of demand between electric and nonelectric energy; and
4. miscellaneous unidentified differences.

The first factor is quantitatively insignificant (see Table D.2), largely because the price projections made by ETA and Nordhaus are not materially different. Differences in conversion loss projections (numbers 2 and 3 above) are quite important, however, totalling 27.6 Quads, or almost the entire remaining difference.

The difference in synthetic fuel conversion losses arises because ETA projects much larger synthetic fuel production than Nordhaus. Here, differences in price projections between the models do seem to play an important role. Thus, ETA projects oil prices in the Base Case (see Table III.16) in 2010 to be well above the $\$3.70/10^6$ BTU price that the MRG assumes will trigger the use of synthetic fuels; by contrast the Nordhaus oil price projection for 2010 is barely above $\$3/10^6$ BTU. This difference in price projections, which is attributable not only to the price and income elasticity differences between the models, but to others as well, can therefore lead to differences in fuel mixes. These, in turn, can produce differences in projections of aggregate primary energy given the way in which aggregate BTU's are accounted for and measured.

A somewhat different explanation lies at the heart of the difference in electricity-generation conversion losses projected by the models. It was shown in Section III.6 that the cross-elasticity of electric energy demand with respect to the price of nonelectric energy is lower in the Nordhaus model than in the ETA model. This does not affect the secondary

TABLE D.1 Contribution of Differences in Income and Price Elasticities Between ETA and Nordhaus to Differences in ETA and Nordhaus Base-Case Projections of Primary Energy Consumption

	Primary Energy Consumption (Quads) in 2010
ETA Projection:	178.9
Nordhaus Projection:	129.8
Difference in the Projections:	49.1

Source of Difference:	Quads (unless otherwise indicated)
I. *Income elasticities*	
Nordhaus primary energy consumption	129.8
Conversion losses	-30.8
Nordhaus secondary energy consumption	99.0
ETA provision for direct burning of coal	-9.0
"Corrected" Nordhaus secondary energy consumption	90.0
Difference between ETA and Nordhaus income elasticities (1.0 - 0.89)	×0.11
Difference in secondary energy-consumption projections due to differing income elasticities	9.9
Ratio of primary to secondary energy required in Nordhaus	×1.3
Difference in primary energy-consumption projections due to differing income elasticities	13.0
II. *Price elasticities*	
ETA secondary energy-consumption assuming Nordhaus price projections	122.7
ETA provision for direct burning of coal	-9.0
"Corrected" ETA secondary energy consumption	113.7
Nordhaus secondary energy consumption corrected for income elasticity difference (99.0 × 1.11)	-108.9
Difference in secondary energy-consumption projections due to differing price elasticities	4.8
Ratio of primary to secondary energy required in Nordhaus	1.3
Difference in primary energy-assumption projections due to differing price elasticities	6.3
III. *Total difference in primary energy projections due to differences in income and price elasticities*	19.3

TABLE D.2 Contribution of Nonelasticity Factors to Differences in ETA and Nordhaus Base-Case Projections of Primary Energy Consumption

	Primary Energy Consumption in 2010
From Table D.1:	
Difference in the Projections:	49.1
Difference due to Elasticity Differences:	19.3
Remainder due to Nonelasticity Differences:	29.8
Source of Difference:	Quads (unless otherwise indicated)
I. *Price projections*	
ETA secondary energy consumption (corrected for 9 Quads of directly burned coal)	113.5
ETA secondary energy consumption assuming Nordhaus price projections (corrected for 9 Quads of directly burned coal)	-113.7
Difference in secondary energy-consumption projections due to higher prices projected by ETA	-0.2
Ratio of primary to secondary energy required in Nordhaus	×1.3
Difference in primary energy-consumption projections due to higher prices projected by ETA	-0.3
II. *Higher synthetic fuel conversion losses in ETA than in Nordhaus*	18.1
III. *Differing electricity-generation conversion losses due to differing cross-elasticities of demand between electric and nonelectric energy*	
ETA projection of electricity consumption (in Quads)	21.9
ETA projection of electricity consumption corrected for income elasticity difference with Nordhaus and assuming Nordhaus price projections	-15.7
"Corrected" difference in electricity consumption (expressed in Quads of primary energy) projections between ETA and Nordhaus	6.2
Approximate ratio of conversion losses to electricity consumption	×1.8
Difference in conversion losses due to differing electricity consumption projections	11.0
Difference in conversion losses due to differing price elasticities	-1.5
Difference in primary energy projections due to differing cross-elasticities of demand between electric and nonelectric energy	9.5
IV. *Miscellaneous*	2.5
Total "other"	29.8

consumption projection, but leads to much higher electricity-conversion loss projections and hence higher primary energy-consumption projections made by ETA. Table D.2 demonstrates that, after the direct elasticity effects are subtracted, the difference in cross-elasticities of demand contributes 9.5 Quads to the overall difference in the primary energy projection.

The relatively low cross-price elasticity between electric and non-electric energy in the Nordhaus model in 2010 is, in large part, due to its relatively high assumed cross-price elasticity between coal and other energy fuels in the industrial sector. In fact, by 2030 it projects about 30 Quads of direct coal use by industry, whereas in the ETA model direct coal use is limited to 9 Quads in 2010. This also explains why ETA projects much higher synthetic fuels production in 2010 than Nordhaus.

In summary, it has been shown that of the total difference in the Base-Case projections of primary energy consumption of 49.1 Quads, 13.0 Quads are due to differing income elasticity assumptions, 6.3 Quads to differing aggregate price elasticity assumptions, and 27.6 to differing assumptions about the cross-price elasticities between specific energy fuels, with the remaining 2.5 Quads being attributable to differences in other assumptions. Section III.11 presents the results of similar calculations for the other scenarios.

APPENDIX E: MRG ACCOUNTING CONVENTIONS

Energy Production	
Coal, natural gas, oil, and NGL	Reported at BTU content, net of exports and imports
Nuclear	Reported at 10,000 BTU/kWh
Solar electric	Reported at 10,000 BTU/kWh
Solar heating and cooling	Reports energy input at 3.2 times BTU content of the end use. This assumes 100 percent efficiency for replacement of delivered electricity and accounts for 9 percent transmission loss in electric uses.
Shale oil	Reported as equivalent of crude oil replaced (at 5.8×10^6 BTU/barrel)
Hydro, geothermal, bioconversion, and nonelectric AES	Nonfossil replacements for electricity are reported at 10,000 BTU/kWh. Replacements for nonelectric uses are reported at the BTU content of the fossil fuel replaced.
Energy Consumption	Energy production and oil and gas imports
Prices (all in 1975 dollars)	
Coal	$\$/10^6$ BTU, delivered to utilities
Natural gas	$/MCF, delivered to industrial consumers
Oil	$/bbl, crude oil delivered to refiners
Electricity	$/1,000 kWh; busbar (shadow) price for ETA and Nordhaus, delivered industrial price for SRI, and delivered consumer's price for DESOM
Uranium	$/lb U_3O_8, figures for "minimum acceptable price" assumed by MRG to be 50 percent higher than the "forward cost" concept that is normally used in ERDA documents

APPENDIX F

F.1 RESPONDENTS TO THE SECOND-ROUND POLL

NOTE: The mean of the responses to the second-round poll from the MRG members and consultants listed below was employed in the R&D decision analyses reported in Sections IV.3 and IV.4.

Clark Bullard, University of Illinois
Edward Cazalet, Private Consultant
Harry Davitian, Brookhaven National Laboratory
Kenneth Hoffman, Brookhaven National Laboratory
William Hogan, Stanford University
Lester Lave, Carnegie-Mellon University
Robert Litan, Council of Economic Advisors
Alan Manne, Electric Power Research Institute
William Nordhaus, Council of Economic Advisors

JUDGMENTAL PROBABILITY ESTIMATE QUESTIONNAIRE

Probability Estimate

Resolved by 1986

1. Energy demand relationships

1a. Demand growth (1975 to 2010) *at constant 1970 level prices,* determined by $E = AP^{\alpha} Y^{\delta}$ and

$$\frac{\dot{E}}{E} = \frac{\dot{A}}{A} + \delta \frac{\dot{Y}}{Y}$$

		$\frac{\dot{A}}{A}$	δ	$\frac{\dot{Y}}{Y}$	$\frac{\dot{E}}{E}$	Probability Estimate
(exclusive of voluntary conservation, assuming		-.3	.8	1.8	1.14	________
Series II population growth)	B*	-.3	1	3.4	3.1	________
(Rates are in percent per year)		0	1	4.7	4.7	________

1b. Voluntary or imposed non-price-induced Conservation Effect

B* 10% on 2010 demand (~ −.3% per year impact on demand growth) ________

None ________

1c. Long run price elasticity of demand in vicinity of current aggregate energy prices

i. electric = .1 ________

.3 ________

.6 ________

ii. nonelectric (primary resource) = .1 ________

.3 ________

.6 ________

(No base value specified)

1d. Population growth .5% ________

.8% (Series II) ________

1.2% ________

2. Coal and shale Constraints

B none ________

50 quads per year (plus 12 quads of shale) ________

*B - MRG Base-Case assumption.

		Probability Estimate
3a. Cost of nonelectric Alternative Energy System Delivered to final consumers	$\$/10^6$ Btu Oil equivalent	
	2	____
	B* 5	____
	8	____
3b. Availability of imported oil and gas, 1990-2010, fraction of domestic oil and gas annual consumption	10%	____
	33%	____
	70% or 20 MMbpd whichever is lower	____
4. Public Acceptance, Safety and Technical Feasibility (determination by 1986)†		
	No nuclear	____
	LWR only, without Pu recycle (no advanced reactors)	____
	B Advanced Converter (ACR) and LWR (with Pu recycle)	____
	LMFBR and LWR, no ACR	____
	LMFBR and ACR	____

Resolved by 1996

5.† Uranium Resources as function of selling price (not "forward costs").

	$/lb. <45	$/lb. 45-150	$/lb. 150-225	Probability Estimate
	2.2	2.8	3.3	____
B	3.7	5.5	4.8	____
	6.5	7	6	____

*B = MRG Base-Case assumption.

†The definition of the energy demand growth and uranium resource levels as used in the R&D decision analysis were revised after the questionnaire was circulated. Using the judgmental probability estimates from the second round and the original resource levels shown here, a cumulative probability distribution curve was drawn from which probability estimates for the revised resource levels were derived.

Resolved after 1996

6. Capital cost differential between LWR and LMFBR

		Probability Estimate
B	0	
	25%	
	50%	

7. Capital cost differential between LWR and ACR

		Probability Estimate
	0	
B	10%	
	25%	

8. Cost of Electricity Producing Alternative Energy Systems for 6GW implementation by 2000

8a. Solar-electric

	Capital Cost	Probability Estimate
	1.25 x LWR	
	2 x LWR	
B	2.6 x LWR ($1,730/KW)*	
	> 5 x LWR	

8b. Fusion, 6GW by 2020

Capital Cost	Probability Estimate
1 x LWR	
2 x LWR	
3 x LWR	
5 x LWR	
∞	

*Base-Case solar capital cost.

F.2 R&D COSTS USED IN ETA DECISION ANALYSIS

Table F.1 presents details on the costs for the R&D strategies of Figure IV.2. In this table, the costs are undiscounted, and expressed in 1975 dollars. For discounting purposes, it is essential that costs be broken down over time. ERDA has undertaken such a breakdown, but has not yet released the results of its study. Consequently, Table F.1 is based on considerable guesswork in order to supplement the time stream of costs *implied* by several ERDA documents.*

The basic assumptions used in calculating the costs of alternative R&D strategies are:

1. *Undiscounted* costs are independent of program phasing. That is, at zero discount rate, the program costs are $12.8 billion regardless of when the breeder becomes available. This assumption is consistent with a declining *discounted* R&D cost assumption.
2. Base program timing is unaffected by changes in the timing of the CRBR, PLBR, and CBR-1 projects.†

Figure F.1 presents the costs of the alternative R&D strategies discounted at 6 percent per year to 1975.‡ If we choose the upper path

*The base program costs in Table F.1 were derived from Table 11.2-3 of ERDA's proposed final environmental statement (PFES) on the LMFBR. See "Final Environmental Statement Liquid Fast Metal Breeder Reactor Program," ERDA-1, January 1975. There, Base Program Costs = Total LMFBR & Support

- (CRBR R&D Costs + Cooperative Projects)
= $6.8 billions (undiscounted 1975 dollars)

The $2.4 billions base program costs for the period 1975-1980 were calculated directly from Table 11.2-3. The remaining $4.4 billion were spread out over the period 1980-2020 in such a way as to justify ERDA's discounted total in the PFES (p. 11.2-3).

From August 1972 to September 1974, ERDA's estimate of CRBR costs rose from $.5 billions to $1.2 billions (1974 dollars). A recent General Accounting Office report states that the last figure was more speculative than it should have been, and hints that costs may increase even further. See "Costs and Schedule Estimates for the Nation's First Liquid Metal Fast Breeder Reactor Demonstration Power Plant," RED-75-358, May, 1975. To be on the safe side, CRBR costs are placed at $2 billion (1975 dollars).

Similarly, PLBR and CBR-1 costs are also placed at $2 billion. Each of these reactors would be a first-of-a-kind machine. Costs should not be significantly affected by the increase in unit size--from 350 MW for CRBR to 1,200 MW for CBR-1.

†The base program refers to those parts of the program that are not included in the description of the demonstration program (e.g., reactor physics, advanced fuels development, etc.).

‡Costs for the period 1975-1980 are discounted as of 1977, 1980-1985 as of 1982, etc.

TABLE F.1 R&D Cost Projections 1975-2020 (Billions of Undiscounted 1975 Dollars)

	1975-1980	1980-1985	1985-1990	1990-1995	1995-2000	2000-2005	2005-2010	2010-2015	2015-2020	Undiscounted Total, 1975-2020
Base Program	2.4	1.0	0.9	0.7	0.6	0.4	0.3	0.3	0.2	6.8
Demonstration Program										
Concurrent development										
CRBR-1983	1.0	1.0								2.0
PLBR-1988		1.0	1.0							2.0
CBR-1 1995			1.0	1.0						2.0
Concurrent development										
CRBR-1983	1.0	1.0								2.0
PLBR-1988		1.0	1.0							2.0
CBR-1 2005					1.0	1.0				2.0
Sequential development										
CRBR-1983	1.0	1.0								2.0
PLBR-1993			1.0	1.0						2.0
CBR-1 2005					1.0	1.0				2.0
Wait										
CRBR-1993			1.0	1.0						2.0
PLBR-1998				1.0	1.0					2.0
CBR-1 2005					1.0	1.0				2.0

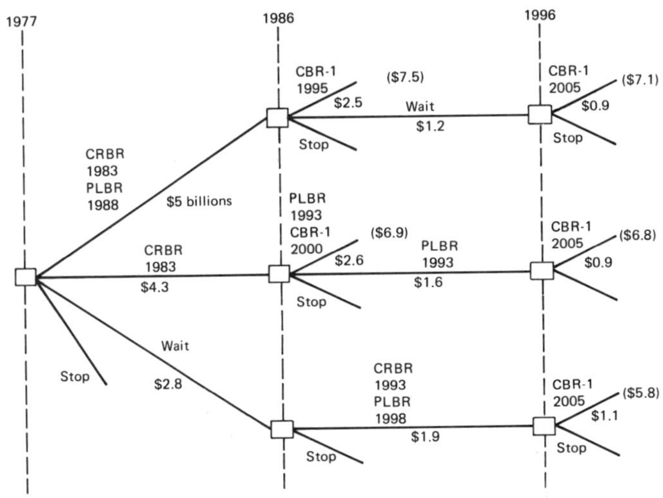

FIGURE F.1 Decision nodes and discounted costs of alternative RD&D strategies. NOTE: Number along each arc denotes present value of RD&D costs, discounting at 6 percent in "real terms" (billions of 1975 dollars). Number in parentheses denotes total costs incurred along each successive arc leading to that terminal node.

and the 1986 decision is CBR-1 in 1995, then the program's discounted costs are $7.5 billion. However, if the concurrent development option is followed and the CBR-1 is then delayed until 2005, discounted costs drop to $7.1 billion. The cost savings from later availability are the result of discounting the costs of CBR-1 from a point further out in time.

From Table F.1, we see that with the sequential development strategy, the undiscounted costs of PLBR are also pushed further into the future. Consequently, if we knew that the CBR-1 would not be needed until 2005, the sequential would be clearly preferable to the concurrent strategy.

If the "wait" option is taken, CBR-1 would be delayed until 2005--at the earliest. This path has the lowest discounted costs of the three 2005 availability paths. Hence, although sequential development is less costly than concurrent development (in discounted dollars), the greatest cost savings would result from delaying the whole program as long as possible.

The R&D costs savings from a 10-year delay in breeder availability are:

Cost of Concurrent Development		Costs of Waiting and Then Concurrent Development
CRBR-1983, PLBR-1988	-	CRBR-1993, PLBR-1998
7.5 billion	-	$5.8 billion
	= $1.7 billion	